Cryospheric Systems: Glaciers and Permafrost

Special Publication reviewing procedures

The Society makes every effort to ensure that the scientific and production quality of its books matches that of its journals. Since 1997, all book proposals have been refereed by specialist reviewers as well as by the Society's Books Editorial Committee. If the referees identify weaknesses in the proposal, these must be addressed before the proposal is accepted.

Once the book is accepted, the Society has a team of Book Editors (listed above) who ensure that the volume editors follow strict guidelines on refereeing and quality control. We insist that individual papers can only be accepted after satisfactory review by two independent referees. The questions on the review forms are similar to those for *Journal of the Geological Society*. The referees' forms and comments must be available to the Society's Book Editors on request.

Although many of the books result from meetings, the editors are expected to commission papers that were not presented at the meeting to ensure that the book provides a balanced coverage of the subject. Being accepted for presentation at the meeting does not guarantee inclusion in the book.

Geological Society Special Publications are included in the ISI Index of Scientific Book Contents, but they do not have an impact factor, the latter being applicable only to journals.

More information about submitting a proposal and producing a Special Publication can be found on the Society's web site: www.geolsoc.org.uk/bookshop.

It is recommended that reference to all or part of this book should be made in one of the following ways:

HARRIS, C. & MURTON, J. B. (eds) 2005. *Cryospheric Systems: Glaciers and Permafrost*. Geological Society, London, Special Publications, **242**.

HUMLUM, O. 2005. Holocene permafrost aggradation in Svalbard. *In*: HARRIS, C. & MURTON, J. B. (eds) 2005. *Cryospheric Systems: Glaciers and Permafrost*. Geological Society, London, Special Publications, **242**, **119–130**.

GEOLOGICAL SOCIETY SPECIAL PUBLICATION NO. 242

Cryospheric Systems: Glaciers and Permafrost

EDITED BY

C. HARRIS

Cardiff University, UK

and

J. B. MURTON

University of Sussex, UK

2005
Published by
The Geological Society
London

THE GEOLOGICAL SOCIETY

The Geological Society of London (GSL) was founded in 1807. It is the oldest national geological society in the world and the largest in Europe. It was incorporated under Royal Charter in 1825 and is Registered Charity 210161.

The Society is the UK national learned and professional society for geology with a worldwide Fellowship (FGS) of 9000. The Society has the power to confer Chartered status on suitably qualified Fellows, and about 2000 of the Fellowship carry the title (CGeol). Chartered Geologists may also obtain the equivalent European title, European Geologist (EurGeol). One fifth of the Society's fellowship resides outside the UK. To find out more about the Society, log on to www.geolsoc.org.uk.

The Geological Society Publishing House (Bath, UK) produces the Society's international journals and books, and acts as European distributor for selected publications of the American Association of Petroleum Geologists (AAPG), the American Geological Institute (AGI), the Indonesian Petroleum Association (IPA), the Geological Society of America (GSA), the Society for Sedimentary Geology (SEPM) and the Geologists' Association (GA). Joint marketing agreements ensure that GSL Fellows may purchase these societies' publications at a discount. The Society's online bookshop (accessible from www.geolsoc.org.uk) offers secure book purchasing with your credit or debit card.

To find out about joining the Society and benefiting from substantial discounts on publications of GSL and other societies worldwide, consult www.geolsoc.org.uk, or contact the Fellowship Department at: The Geological Society, Burlington House, Piccadilly, London W1J 0BG: Tel. +44 (0)20 7434 9944; Fax +44 (0)20 7439 8975; E-mail: enquiries@geolsoc.org.uk.

For information about the Society's meetings, consult *Events* on www.geolsoc.org.uk. To find out more about the Society's Corporate Affiliates Scheme, write to enquiries@geolsoc.org.uk.

Published by The Geological Society from:
The Geological Society Publishing House
Unit 7, Brassmill Enterprise Centre
Brassmill Lane
Bath BA1 3JN, UK

(*Orders*: Tel. +44 (0)1225 445046
        Fax +44 (0)1225 442836)
Online bookshop: http://bookshop.geolsoc.org.uk

The publishers make no representation, express or implied, with regard to the accuracy of the information contained in this book and cannot accept any legal responsibility for any errors or omissions that may be made.

© The Geological Society of London 2005. All rights reserved. No reproduction, copy or transmission of this publication may be made without written permission. No paragraph of this publication may be reproduced, copied or transmitted save with the provisions of the Copyright Licensing Agency, 90 Tottenham Court Road, London W1P 9HE. Users registered with the Copyright Clearance Center, 27 Congress Street, Salem, MA 01970, USA: the item-fee code for this publication is 0305-8719/05/$15.00.

British Library Cataloguing in Publication Data

A catalogue record for this book is available from the British Library.

ISBN 1-86239-175-0
Project managed by Techset Composition, Salisbury, UK
Printed by The Cromwell Press, Wiltshire, UK.

Distributors

USA
  AAPG Bookstore
  PO Box 979
  Tulsa
  OK 74101-0979
  USA
Orders: Tel. +1 918 584-2555
        Fax +1 918 560-2652
        E-mail bookstore@aapg.org

India
  Affiliated East-West Press Private Ltd
  Marketing Division
  G-1/16 Ansari Road, Darya Ganj
  New Delhi 110 002
  India
Orders: Tel. +91 11 2327-9113/2326-4180
        Fax +91 11 2326-0538
        E-mail affiliat@vsnl.com

Japan
  Kanda Book Trading Company
  Cityhouse Tama 204
  Tsurumaki 1-3-10
  Tama-shi, Tokyo 206-0034
  Japan
Orders: Tel. +81 (0)423 57-7650
        Fax +81 (0)423 57-7651
        E-mail geokanda@ma.kcom.ne.jp

Contents

Preface

The interaction between glacial and periglacial environments may strongly influence processes of erosion and deposition, and in addition, regional groundwater and surface water flows may be severely modified. During periods of climate change, the coupling of glacier and permafrost thermal regimes leads to complex but highly significant process responses. The geographical zones of interaction between permafrost and glaciers may differ markedly between the polar regions and the mid-latitude mountains, but in both cases understanding these zones may prove critical to anticipating impacts of future climate warming. There is, therefore, an enormous potential for interdisciplinary research between glaciologists and geocryologists. With this in mind, a conference entitled "Cryospheric Systems: Glaciers and Permafrost", sponsored by the British Geomorphological Research Group and the Quaternary Research Association, was held at the Geological Society of London in January 2003. Selected conference contributions are published in the present volume, including papers concerned with processes occurring at the permafrost-glacier interface as well as papers concerned solely with either glacial systems or periglacial systems. The overall purpose of bringing this work together is to highlight the interactions between glaciers and permafrost in order to stimulate further cross-disciplinary research between glaciology and geocryology. The editors wish to thank the authors for their diligence in submission of manuscripts, and also to gratefully acknowledge the work of the following reviewers:

Colin Ballantyne, Doug Benn, Matthew Bennet, Christopher Burn, Hanne Christiansen, Bernd Etzelmuller, Sean Fitzsimmons, Hugh French, Neil Glasser, Michael Hambrey, Stephan Harrison, Richard Hodgkins, Bryn Hubbard, Ole Humlum, Roy Koerner, Danny McCarroll, Fredrick Nelson, Brice Rea, Jef Vandenburghe, Jeff Warburton and Brian Whalley.

Charles Harris
Julian B. Murton
January 2005

Interactions between glaciers and permafrost: an introduction

CHARLES HARRIS[1] & JULIAN B. MURTON[2]

[1]School of Earth, Ocean and Planetary Sciences, Cardiff University, Cardiff, UK
[2]Department of Geography, University of Sussex, Brighton, UK

Abstract: A consideration of the interactions between glaciers and permafrost is essential to many environmental studies of cold regions. This paper reviews how concepts, field data and experimental studies from glaciology and geocryology can provide a basis for improved understanding of some of the key glacier–permafrost interactions at scales ranging from continental ice sheets to small proglacial streams. Glacitectonic processes are strongly influenced by water pressures beneath subglacial and proglacial permafrost, and by the amount of unfrozen water within the permafrost. Burial of glacier ice and growth of intra-sedimental ice also occur within sub- and proglacial permafrost, and together they produce a complex assemblage of ground ice in glaciated frozen lowlands. In mountain regions, rock glaciers are associated with the presence of ground ice, and represent a landform that straddles the semantic fence separating glacial features from permafrost features. In proglacial and ice-marginal environments, geomorphological activity reflects the combined effects of glacially- and periglacially-conditioned processes operating synchronously in adjacent areas, or in succession; such activity includes the transport of sediment in glacierized catchments and the calving of glacial ice in ice-marginal lakes. Interaction between permafrost and glacial phenomena depends largely on their proximity: where permafrost occurs close to glaciers, the thermal regime of the active layer is influenced in part by the surface covering of adjacent glacial ice through its effect on albedo and ground heat flux. Glacier–permafrost interactions are particularly important in recently deglaciated terrain, where permafrost may be aggrading. This 'paraglacial' zone often shows rapid geomorphological change. Thus, glacier–permafrost interactions are complex and occur over a wide range of temporal and spatial scales.

The cryosphere (derived from the Greek *kryos*, meaning 'cold') refers to those parts of the Earth's surface where water is dominantly frozen, and includes areas with snow cover, sea ice, glaciers and permafrost. The monitoring of cryospheric parameters that currently contribute to the Global Climate Observation System includes the temperature of cold firn, glacier size and mass balance, timing of lake and river freeze-up and break-up, permafrost temperature, active-layer thickness, and snow cover area and water equivalent (Dyurgerov & Meier 1997; Vinnikov *et al.* 1999; Armstrong & Brodzik 2001; Harris *et al.* 2001). Complex interactions amongst these parameters and their feedback effects on geomorphological processes (French 1996; Benn & Evans 1998), the behaviour of cold-based glaciers and ice sheets (Cutler *et al.* 2000; Waller 2001), the global climate system (Etkin & Agnew 1992; IPCC 2001) and the long-term evolution of cold landscapes (Fulton 1989; Ballantyne & Harris 1994) underline the importance of studying the cryosphere from an interdisciplinary perspective. This book con-

siders some of the key interactions between glaciers and permafrost.

In a geological context, glaciers and glacial processes have generated more scientific literature than permafrost and related phenomena, almost certainly because mid- and high-latitude landscapes often show clear evidence for glaciation, both in the form of erosional and depositional landforms. Research into permafrost and frozen ground phenomena began with early observations by Russian scientists and explorers in Siberia (Yershov 1998), but has since lagged behind studies of glaciers and glacial processes. Initially the term *geocryology* (Russian *geokriologiya*) was used to describe the science of the cryosphere, including both frozen ground and glaciers (Fyodorov & Ivanov 1974). However, a dichotomy between the study of glaciers (glaciology) and frozen ground resulted from the introduction of the term *periglacial* by Łoziński (1909, 1912) to describe the cold-climate conditions in the zone adjacent to but beyond the Pleistocene glaciers, and the distinctive frost-related geomorphological pro-

From: HARRIS, C. & MURTON, J. B. (eds) 2005. *Cryospheric Systems: Glaciers and Permafrost.*
Geological Society, London, Special Publications, **242**, 1–9.
0305-8719/05/$15.00 © The Geological Society of London 2005.

cesses that prevailed there (Haeberli 2005). Through the twentieth century, the field of geocryology has come to exclude glaciers and to focus on frozen ground phenomena (Washburn 1979; Yershov 1998). As Haeberli points out, the geological and geomorphological processes at the interface between glaciers and permafrost have therefore received less attention than they warrant, and the influence of one on the other has been largely neglected. The present chapter discusses the glacier–permafrost interactions that underpin (1) glacitectonic processes, (2) ground-ice development, (3) rock glaciers, (4) proglacial and ice-marginal processes and (5) permafrost and related processes.

Glacitectonic processes

Study of glacitectonic structures in northern Germany led to the early recognition that, where glacial margins advance across permafrost, stress generated by the advancing glacier may be transferred over considerable distances (Kozarski 1959). Kozarski argued that the large Chodziez moraine complex, in Poland, resulted from glacitectonic thrusting of shallow permafrost over a substrate of soft, unfrozen Pliocene clay. However, permafrost may not always be necessary for glacitectonic deformation, for example, if the substrate contains strata sufficiently weak to allow rapid shearing (Aber et al. 1989), or if the ice margin advances across a topographic obstruction within the pre-existing terrain (Schott 1933). The influence of unfrozen water on glacitectonic deformation is of particular importance, firstly along the base of the permafrost and secondly within the permafrost itself.

The shearing resistance along a basal plane of décollement is a function of the coefficient of friction and effective normal stress on the plane. Mathews & Mackay (1960) argued that high porewater pressures beneath subglacial permafrost facilitate glacial thrusting by reducing the shear strength of unfrozen soil. Boulton et al. (1995) modelled subglacial water pressures and effective stress, in a transect through Denmark and the Netherlands, during the Saalian advance of the European ice sheet. Their modelling suggested that a continuous permafrost cover in the proglacial zone that was coupled to the ice margin would generate overpressuring in groundwater beneath proglacial permafrost, thus reducing effective stresses. In certain cases, pore pressures would be sufficient to induce a zone of negative effective stress (pore pressures exceeding overburden pressure) in the proglacial zone adjacent to the glacier

and in a narrow zone beneath the ice margin. Boulton & Caban (1995) demonstrated that this marginal overpressuring is conducive to the formation of large- and intermediate-scale diapiric structures (termed 'extrusion moraines') and large push moraines, formed as a result of low frictional resistance along the décollement plane at the permafrost base, and stiffening of the mobilized sediments by permafrost. Where the frozen foreland is traversed by large river systems underlain by through-going taliks – for example, beneath the rivers Rhine, Saale and Elbe – efflux of pressurized groundwater into the talik and absence of permafrost provide a zone of low resistance, facilitating thrusting of the frozen proglacial permafrost plate and formation of large push moraine complexes.

Glacitectonic deformation that occurs within permafrost is strongly influenced by the amount of unfrozen water present in frozen soil. Such water reduces the strength of bonds between ice crystals and soil particles, thus lowering the cohesion of the ice matrix. The amount of unfrozen water in ice-cemented soil depends in large part on temperature and particle size. Sand usually contains negligible amounts of unfrozen water at temperatures marginally below $0°C$, whereas clay contains significant amounts at temperatures of a few degrees or more below $0°C$, the amounts in all soils decreasing with temperature (Williams & Smith 1989). As a result, frozen, fine-grained soils generally have much lower yield stresses than frozen coarse-grained soils, and at temperatures between 0 and $-2°C$ are more susceptible to creep deformation than is pure ice (Johnston et al. 1981). Such *warm* conditions (i.e., close to the pressure melting point) result in a reduction in the yield stress of both the ice and frozen sediment and thereby promote pervasive creep deformation (Vialov 1965; Mellor & Testa 1969). This type of easily deformable permafrost has been termed 'plastic-frozen ground' (Tsytovich 1975), as opposed to 'hard-frozen ground', which contains very little unfrozen water and is comparatively rigid.

Cryostratigraphical studies of Pleistocene permafrost in glaciated lowlands provide opportunities to examine former glacier–bed interactions beneath cold-based ice. Previous studies in western Arctic Canada and northwest Siberia have provided important evidence for glacier-ice thrusting of permafrost near the northwest limit of the Laurentide Ice Sheet (Mathews & Mackay 1960; Mackay et al. 1972), and for a thick deformable bed of permafrost beneath the southern Kara Ice Sheet (Astakhov et al. 1996). Recent study has suggested that deformation of ice-rich permafrost beneath the northwest margin of the Laurentide Ice Sheet was (a) mainly ductile in character, (b) involved relatively competent units of frozen sand and ice clasts within a more rapidly deforming shear zone of partially frozen diamicton (glacitectonite), and (c) accompanied by sub-marginal erosion of pre-existing permafrost (Murton *et al.* 2005a). Logically, the next stage of research is to integrate cryostratigraphical, glacitectonic, geophysical and remote-sensing data over regional scales in order to reconstruct the dynamic behaviour of the margins of large ice lobes during the late Pleistocene. Such data are vital for testing and refining models of interactions between ice lobes and permafrost (e.g., Cutler *et al.* 2000).

Several authors, including Hambrey *et al.* (1996) and Boulton *et al.* (1999), have discussed the role of proglacial permafrost in the formation of smaller-scale push moraines by surging glaciers, particularly in the arctic islands of Svalbard. A detailed discussion of mechanisms of permafrost–glacier interaction, including its role in the formation of push moraines is presented in this volume by Etzelmüller & Hagen. Waller & Tuckwell (2005) also describe glacitectonic structures, in this case at the Leverett Glacier, West Greenland, and argue for a two-phase development. Bennett *et al.* (2005) suggest that a winter glacier surge in Iceland caused ice advance across seasonally frozen glaciofluvial sediments that were sufficiently stiff to transmit the lateral stresses and cause formation of a distinctive thrust moraine sequence.

Ground-ice development

Ground ice can be defined as ice that occurs within frozen or partly frozen ground, irrespective of the form of occurrence or origin of the ice. Glacier–permafrost interactions are an important influence on ground-ice development because glaciers often supply large amounts of ice and meltwater to subglacial or proglacial permafrost. Nowhere are these interactions more important than in regions where ice sheets advance and retreat across lowlands underlain by thick, frozen sequences of unconsolidated sediments, as recorded in the permafrost of western Arctic Canada and northwest Siberia. Such permafrost contains a complex assemblage of *buried* ice and *intrasedimental* ice.

Buried glacier ice is of two types: firn-derived glacier ice (*firnified glacier ice*) and basal glacier ice (*basal ice*). Burial results where (1) sediments are deposited on top of the ice by meltwater, mass movement, lacustrine and aeolian processes (e.g., Moorman & Michel 2000), (2) debris melts out from basal ice, forming an insulating blanket of supraglacial melt-out till or (3) glacial thrusting and shearing emplaces frozen sediment above the ice. If the ice margin lies within the permafrost zone and if the debris cover is thicker than the active layer, the buried ice can be preserved indefinitely in permafrost and becomes part of the permafrost system. In western Arctic Canada, buried ice from the Laurentide Ice Sheet has been preserved for more than 10 ka within large moraine belts, hummocky till and glaciofluvial deposits (French & Harry 1990; Sharpe 1992; Dredge *et al.* 1999; St-Onge & McMartin 1999; Dyke & Savelle 2000; Murton *et al.* 2005b); in northwestern Siberia buried ice from the Barents–Kara Ice Sheet has been preserved for c. 80–90 ka (Kaplyanskaya & Tarnogradskiy 1986; Astakhov & Isayeva 1988; Svendsen *et al.* 2004); and in southern Victoria Land, Antarctica, buried ice has been preserved since the Miocene (Marchant *et al.* 2002). On a smaller scale, buried glacier ice commonly develops in arctic and alpine mountain environments, where supraglacial debris buries the ice-marginal zones of many glaciers (Etzelmüller & Hagen 2005), some of which develop into debris-covered rock glaciers. If the ice-marginal zone is too warm for permafrost to occur, geothermal and surface heat fluxes are likely to cause gradual decay of the ice core. Such ice-cored moraines are not then part of a permafrost domain (Haeberli 2005). The complex interface between polythermal glaciers and continuous permafrost is explored by Moorman (2005) in the specific context of meltwater flows from the glacier into the proglacial permafrost zone.

Intrasedimental ice develops where water freezes or water vapour sublimates within soil, sediment or bedrock. It develops in proglacial permafrost in two situations: (1) where pressurized glacial meltwater freezes onto the base of the permafrost or is injected up into it; or (2) where permafrost re-aggrades through sediments that are subglacially thawed by the insulating effects of a thick body of glacier ice, and then, following glacier retreat, subaerially frozen upon re-exposure to cold winter air temperatures. In both cases, the overall effect of the glacier or ice sheet, except where the ice is predominantly cold-based or where its substrate has a very low permeability, is to predispose the permafrost to becoming *ice-rich* (i.e., where the ice volume exceeds the pore volume of the ground in an unfrozen state). Besides pore ice (ice cement), the most important types of proglacial intrasedimental ice, in terms of glacier–permafrost interactions, are segregated ice and intrusive ice. These occur as ground-ice bodies that vary from ice lenses and veins of millimetre size or less to tabular bodies of massive ice tens of metres thick and hundreds of metres or more in horizontal extent.

An elegant model for the growth of massive intrasedimental ice has been proposed by Mackay (1971, 1989) and Mackay & Dallimore (1992). The model involves the growth of segregated to intrusive ice as permafrost aggrades downwards through ground subject to high pore water pressures. The high pressures arise from (1) pore-water expulsion during freezing of saturated sand, as occurs during growth of hydrostatic pingos (Mackay 1998), but on a larger scale and/or (2) a hydraulic gradient generated, for example, in subglacial meltwater driven towards aggrading permafrost in the proglacial zone of an ice sheet (Rampton 1974), as modelled by Boulton et al. (1995) and Boulton & Caban (1995). Mackay suggested that pressurized groundwater nourishes a lens of massive ice growing at or near the contact between silt–clay (frost-susceptible) sediment and underlying sand. The ice type depends on the water pressure: segregated ice forms where the water pressure is sufficient to drive unfrozen water to the freezing front but insufficient to lift the frozen overburden, whereas intrusive ice forms where very high water pressures lift the overburden, as in the case of 'pulsating' pingos, whose ice core rests partly on a lens of highly pressurized water (Mackay 1977). Fluctuating water pressures across the overburden uplift threshold result in interlayering of segregated and intrusive ice. Another important geological consequence of groundwater pressures in excess of lithostatic is the hydraulic fracturing of ice-marginal permafrost and the injection of water that on freezing forms dykes and sills of intrusive ice (Mackay 1989; Mackay & Dallimore 1992).

Intrasedimental ice can also form subglacially and contribute to the basal ice layer of glaciers and ice sheets. Pore, segregated and intrusive ice can all freeze onto the bottom of a glacier or ice sheet – regardless of whether the bottom comprises ice or clastic sediment – providing that the freezing front descends beneath the ice body; such ice forms both subglacially and intrasedimentally, for example within preglacial or glacigenic sediments that become incorporated into a basal ice layer.

Given these different ways of producing bodies of massive ice and icy sediments in permafrost, it is not surprising that their interpretation has been subject to debate, as perhaps best illustrated from the Tuktoyaktuk Coastlands of western Arctic Canada, where the ice has been examined in particular detail (Mackay 1971;

Rampton & Mackay 1971; Rampton 1974, 1988, 1991; Dallimore & Wolfe 1988; French & Harry 1990; Mackay & Dallimore 1992; Murton et al. 2005b; Murton 2005). The picture that is emerging from this area, especially from cryostratigraphic studies of glacially deformed permafrost, is that much of the ground ice has developed in response to the dynamic interactions between the northwest margin of the Laurentide Ice Sheet and the permafrost beneath and, later, in front of it. During Marine Isotope Stage 2, the margin of the ice sheet advanced across permafrost, causing, in some areas, deformation of massive segregated ice, erosion of pre-existing ground ice, burial of basal ice by glacier thrusting, and formation of frozen glacitectonites. By analogy with basal ice at the margin of contemporary, sub-polar ice sheets such as that in western Greenland (Knight 1997), the northwest margin of the Laurentide Ice Sheet probably had a thick basal ice layer formed by accretion of both new and existing ice. During deglaciation, pressurized meltwater from the warm-based interior of the ice sheet flowed northwards beneath proglacial permafrost, in places hydraulically fracturing the permafrost and forming dykes and sills of intrusive ice and supplying water to bodies of massive intrasedimental ice. Unlike the older, glacially deformed ground-ice, the ice formed during deglaciation lacks evidence for glacial deformation and it truncates glacitectonic structures. Some postglacial massive intrasedimental ice formed as a result of permafrost aggradation and pore water expulsion from saturated sandy sediments, as modelled by Mackay. Deglaciation was accompanied by significant erosion of Pleistocene sediments by glacial meltwater, and burial of the basal ice continued by a variety of processes, including deposition of aeolian sand and outwash. As the climate continued to warm during the Last Glacial–Interglacial Transition, stagnant masses of the debris-rich basal ice layer were subject to downward melting, burying them beneath supraglacial melt-out till.

Rock glaciers

Rock glaciers comprise lobate or tongue-shaped landforms developed in frozen rock debris as a result of gravitational creep of ground ice within them. Rock glaciers are by definition permafrost features, and therefore part of the permafrost cryozone (e.g., Haeberli 1996). However, much dispute has arisen over their definition because in some cases rock-glacier-like forms extend below cirque glaciers, or former cirque glaciers, and contain buried glacier ice (e.g.,

Humlum 1998). In some respects, this dispute is one of semantics, and reflects the breadth of usage of the term 'rock glacier' (Ballantyne & Harris 1994), but since relict rock glaciers are frequently used to reconstruct mountain permafrost distribution, clarity in definition is essential. Two factors appear critical here, and are discussed in this volume by Haeberli and by Etzelmüller & Hagen: firstly, the thermal regime, and secondly, the influence of glacier ice motion on the dynamics of the 'rock glacier'. Thus, a glacier or glacier margin that is covered in rock debris and contains a slowly melting glacier ice core cannot constitute a rock glacier since it is not part of a permafrost environment. Such a feature is more accurately defined as a degrading ice-cored moraine or rock-covered glacier. Similarly, the rock-covered margin of a cold glacier that is affected by ice motion in the glacier above cannot be classified as a rock glacier since its dynamics are not principally controlled by gravitational creep within the rock-covered marginal zone.

The present distribution of rock glaciers in terms of mean annual air temperature (MAAT) and mean annual precipitation may clearly provide useful information in palaeoclimate reconstruction based on relict features. Haeberli (1982) and Barsch (1996) showed that, in general, for a given MAAT, permafrost, and therefore rock glaciers, occurs in areas of lower precipitation than glaciers (Haeberli 2005, Fig. 1). However, as demonstrated by Humlum (1998), the boundary between each zone is highly diffuse, with true rock glaciers recorded in the same environmental contexts as glaciers. Thus, interpretation of ice-cored, debris-covered terrain in mountain regions requires a clear understanding of both the glacial and the permafrost environments. The complexity of glacier–rock glacier interactions and their relation to climate is well illustrated in the present volume by the paper of Dornbusch.

Proglacial and ice-marginal processes

It is clear that the dominant control of sedimentary processes within proximal glacial meltwater systems is the seasonally modulated melting of ice and snow within the glacier system (e.g., Miall 1983). It is also apparent, however, that when considering the sediment yield and sedimentary processes of many arctic catchments, inputs of sediment and water from the non-glaciated areas have a significant effect (e.g., Marsh & Woo 1981; Woo et al. 2000). Snow melt, break-up of river and lake ice, and permafrost aggradation or degradation influence water discharge; and processes such as proglacial river erosion, slope instability and solifluction influence sediment transport. As a result, permafrost-related processes are likely to be important variables in overall catchment behaviour, as discussed by Irvine-Flynne et al. (2005) for three glacierized arctic catchments.

Rapid landscape adjustment following glacier retreat may include both sediment redistribution and rock-slope instability. The processes whereby such rapid readjustments occur have been termed 'paraglacial' (Church & Ryder 1972). Essentially the paraglacial concept is based on the premise that recently deglaciated terrain is often susceptible to rapid modification by subaerial agents because withdrawal of the glacier may leave rock slopes and sediments in an initially unstable or metastable condition (Ballantyne 2002). A specific example discussed by Mercier & Laffly (2005) is fluvial reworking of sediment, in this case in a coastal area of Spitsbergen, where rapid glacier retreat has occurred since the Little Ice Age.

More specific to the glacier system are mechanisms in the marginal zone affecting discharge of ice, water and sediment. A particular context, and one that has received relatively little attention, is where glaciers terminate in a lake deep enough to cause calving. A case study from Patagonia is presented by Haresign & Warren (2005), who provide a detailed analysis of calving rates and their controls.

Permafrost and related processes

Permafrost and related processes show variable degrees of interaction with glacial phenomena, often dependent on their proximity to masses of glacial ice and hence on exchanges of energy and on transfers of water and ice. The value of monitoring permafrost temperature as an indicator of climate change has already been mentioned. Since short-term temperature cycles at the ground surface are rapidly attenuated with depth in permafrost, the thermal profile provides us with an important record of longer-term changes in surface temperature (e.g., Lachenbruch & Marshall 1986; Harris et al. 2001, 2003). Permafrost development in Svalbard, including recent permafrost aggradation (Humlum 2005), provides an excellent illustration of the interaction between ground temperatures, atmospheric temperatures and surface cover, including glaciers.

Over much longer time scales, geological evidence for extensive permafrost during Quaternary cold stages often includes the presence of ice-wedge pseudomorphs within sedimentary sequences (e.g., Dylik 1966;

Vandenberghe 1983). Interpretation of pseudo-morph geometry and dimensions in terms of the size and form of the original ice wedge, however, is problematic because, during permafrost thaw, deformation of host sediments and mechanisms of wedge casting may lead to pseudomorphs that differ considerably from the original wedges. This problem is addressed by Harris & Murton (2005) through a series of scaled laboratory simulation experiments carried out in a geotechnical centrifuge. The authors show that pseudomorph geometry varies considerably with the granulometry and ice content of the host sediment at the time of thaw, further complicating the palaeoenvironmental interpretation of natural ice-wedge pseudomorphs (Murton & Kolstrup 2003).

In cold regions, many aspects of subaerial denudation are highly sensitive to climate, and changes within sedimentary sequences may provide important evidence for longer-term climate fluctuations. Often a critical element is that of dating the timing of climatically controlled changes in sediment transport and deposition. The final contribution to the present volume, by Kirkbride & Dugmore (2005), is concerned with Late Holocene solifluction history in Iceland, using tephrochronology to provide a temporal framework. Although not necessarily an indicator of permafrost, the rate of solifluction is sensitive to the ground freezing and thawing regime and to variations in moisture supply. Kirkbride & Dugmore (2005) demonstrate that solifluction showed marked variations in activity over the last five millennia, a sequence not exactly paralleling advance and retreat of the local glaciers.

Concluding comments

This collection of papers illustrates in part the range of research presently undertaken within the proglacial periglacial domain. Focus is given to permafrost and its interaction with glacial processes, and to the influence of glaciers on permafrost geomorphological systems, both in contemporary and Pleistocene environments. Since glacier mass balance is a function of ice accumulation and loss, glaciers may exist despite high ablation rates, as long as these are less than accumulation rates. Thus glaciers may be present where permafrost is largely absent, although in many cases at least part of the glacial system is embedded within permafrost terrain. Permafrost may greatly influence the interaction between a glacier and its substrate, especially in ice-marginal zones. Conversely, the presence of a glacier ice cover fundamentally alters the geothermal regime within a permafrost region, so that in high mountains, climate-induced glacier retreat may be associated with permafrost migration into the recently exposed terrain. Where glacier ice becomes buried within proglacial sedimentary sequences, the sensitivity of the landscape to thermal change increases significantly. Thus, this critical interface within the cryosphere should be given high priority, with particular focus on integration of knowledge derived from glaciology and geocryology.

References

ABER, J. S., CROOT, D. G. & FENTON, M. M. 1989. *Glaciotectonic Landforms and Structures.* Glaciology and Quaternary Geology Series, Kluwer Academic, Dordrecht.

ARMSTRONG, R. L. & BRODZIK, M. J. 2001. Recent Northern Hemisphere snow extent: A comparison of data derived from visible and microwave sensors. *Geophysical Research Letters*, **28**, 3673–3676.

ASTAKHOV, V. I. & ISAYEVA, L. L. 1988. The 'Ice Hill'; an example of 'retarded deglaciation' in Siberia. *Quaternary Science Reviews*, **7**, 29–40.

ASTAKHOV, V. I., KAPLYANSKAYA, F. A. & TARNOGRADSKY, V. D. 1996. Pleistocene permafrost of West Siberia as a deformable glacier bed. *Permafrost and Periglacial Processes*, **7**, 165–191.

BALLANTYNE, C. K. 2002. Paraglacial geomorphology. *Quaternary Science Reviews*, **21**, 1935–2017.

BALLANTYNE, C. K. & HARRIS, C. 1994. *The Periglaciation of Great Britain*, Cambridge University Press, Cambridge.

BARSCH, D. 1996. *Rockglaciers: Indicators for the Present and Former Geoecology in High Mountain Environments*, Springer, Berlin.

BENN, D. I. & EVANS, D. J. A. 1998. *Glaciers and Glaciation*, Arnold, London.

BENNETT, M. R., HUDDART, D. & WALLER, R. I. 2005. The interaction of a surging glacier with a seasonally frozen foreland: Hagafellsjökull-Eystri, Iceland. *In*: HARRIS, C. & MURTON, J. B. (eds.): *Cryospheric Systems: Glaciers and Permafrost.* Geological Society, London, Special Publications, **242**, 51–62.

BOULTON, G. S. & CABAN, P. E. 1995. Groundwater flow beneath ice sheets: Part II – its impact on glacier tectonic structures and moraine formation. *Quaternary Science Reviews*, **14**, 563–587.

BOULTON, G. S., CABAN, P. E. & VAN GIJSSEL, K. 1995. Groundwater flow beneath ice sheets: Part I – large scale patterns. *Quaternary Science Reviews*, **14**, 545–562.

BOULTON, G. S., VAN DER MEER, J. J. M., BEETS, D. J., HART, J. K. & RUEGG, G. H. J. 1999. The sedimentary and structural evolution of a recent push moraine complex: Holmstrømbreen, Spitsbergen. *Quaternary Science Reviews*, **18**, 339–371.

CHURCH, M. & RYDER, J. M. 1972. Paraglacial sedimentation; a consideration of fluvial processes conditioned by glaciation. *Bulletin of the Geological Society of America*, **83**, 3059–3072.

CUTLER, P. M., MACAYEAL, D. R., MICKELSON, D. M., PARIZEK, B. R. & COLGAN, P. M. 2000. A numerical investigation of ice-lobe/permafrost interaction around the Southern Laurentide Ice Sheet. *Journal of Glaciology*, **46**, 311–325.

DALLIMORE, S. R. & WOLFE, S. A. 1988. Massive ground ice associated with glaciofluvial sediments, Richards Islands, N.W.T., Canada. *In: Permafrost, Fifth International Conference, Proc.*, Tapir, Trondheim, Vol. 1, 132–137.

DORNBUSCH, U. 2005. Glacier–rock glacier relationships as climatic indicators during the late Quaternary in the Cordillera Ampato, Western Cordillera of southern Peru. *In:* HARRIS, C. & MURTON, J. B. (eds.): *Cryospheric Systems: Glaciers and Permafrost*. Geological Society, London, Special Publications, **242**, 75–82.

DREDGE, L. A., KERR, D. E. & WOLFE, S. A. 1999. Surficial materials and related ground ice conditions, Slave Province, N.W.T., Canada. *Canadian Journal of Earth Sciences*, **36**, 1227–1238.

DYKE, A. S. & SAVELLE, J. M. 2000. Major end moraines of Younger Dryas age on Wollaston Peninsula, Victoria Island, Canadian Arctic: Implications for paleoclimate and for formation of hummocky moraine. *Canadian Journal of Earth Sciences*, **37**, 601–619.

DYURGEROV, M. B. & MEIER, M. F. 1997. Mass balance of mountain and subpolar glaciers: A new global assessment for 1961–1990. *Arctic and Alpine Research*, **29**, 379–391.

DYLIK, J. 1966. Problems of ice-wedge structures and frost-fissure polygons. *Biuletyn Peryglacjalny*, **15**, 241–291.

ETKIN, D. A. & AGNEW, E. 1992. Arctic climate in the future. *In:* WOO, M. K. & GREGOR, D. J. (eds) *Arctic Environment: Past, Present and Future*, McMaster University, Hamilton, 17–34.

ETZELMÜLLER, B. & HAGEN, J. O. 2005. Glacier–permafrost interaction in Arctic and alpine mountain environments with examples from southern Norway and Svalbard. *In:* HARRIS, C. & MURTON, J. B. (eds.): *Cryospheric Systems: Glaciers and Permafrost*. Geological Society, London, Special Publications, **242**, 11–27.

FRENCH, H. M. 1996. *The Periglacial Environment*, Longman, London.

FRENCH, H. M. & HARRY, D. G. 1990. Observations on buried glacier ice and massive segregated ice, western Arctic coast, Canada. *Permafrost and Periglacial Processes*, **1**, 31–43.

FULTON, R. J. (ed.) 1989. *Quaternary Geology of Canada and Greenland*. Geological Survey of Canada, Geology of Canada no. 1 (also Geological Society of America, The Geology of North America, volume K-1).

FYODOROV, J. S. & IVANOV, N. S. 1974. *English–Russian Geocryological Dictionary*, Yakutsk State University, Yakutsk.

HAEBERLI, W. 1982. *Klimarekonstruktionen mit Glescher-Permafrost-Beziehungen*, Baseler Beiträge zur Physiographie, Basel, **4**, 9–17.

HAEBERLI, W. 1996. On the morphodynamics of ice/debris-transport systems in cold mountain areas. *Norsk Geografisk Tidsskrift*, **50**, 3–9.

HAEBERLI, W. 2005. Investigating glacier–permafrost relationships in high-mountain areas: historical background, selected examples and research needs. *In:* HARRIS, C. & MURTON, J. B. (eds.): *Cryospheric Systems: Glaciers and Permafrost*. Geological Society, London, Special Publications, **242**, 29–37.

HAMBREY, M. J., DOWDESWELL, J. A., MURRAY, T. & PORTER, P. R. 1996. Thrusting and debris-entrainment in a surging glacier, Bakaninbreen, Svalbard. *Annals of Glaciology*, **22**, 241–248.

HARESIGN, E. & WARREN, C. R. 2005. Melt rates at calving termini: A study at Glaciar León, Chilean Patagonia. *In:* HARRIS, C. & MURTON, J. B. (eds.): *Cryospheric Systems: Glaciers and Permafrost*. Geological Society, London, Special Publications, **242**, 99–109.

HARRIS, C. & MURTON, J. B. 2005. Experimental simulation of ice-wedge casting: Processes, products and palaeoenvironmental significance. *In:* HARRIS, C. & MURTON, J. B. (eds.): *Cryospheric Systems: Glaciers and Permafrost*. Geological Society, London, Special Publications, **242**, 131–143.

HARRIS, C., HAEBERLI, W., VONDER MÜHLL, D. & KING, L. 2001. Permafrost monitoring in the high mountains of Europe: The PACE project in its global context. *Permafrost and Periglacial Processes*, **12**, 3–11.

HARRIS, C., VONDER MÜHLL, D., ISAKSEN, K., HAEBERLI, W., SOLLID, J. L., KING, L., HOLMLUND, P., DRAMIS, F., GUGLIELMIN, M. & PALACIOS, D. 2003. Warming permafrost in European mountains. *Global and Planetary Change*, **39**, 215–225.

HUMLUM, O. 1998. The climatic significance of rock glaciers. *Permafrost and Periglacial Processes*, **9**, 375–395.

HUMLUM, O. 2005. Holocene permafrost aggradation in Svalbard. *In:* HARRIS, C. & MURTON, J. B. (eds.): *Cryospheric Systems: Glaciers and Permafrost*. Geological Society, London, Special Publications, **242**, 119–130.

IPCC, 2001. *Climate Change 2001: The Scientific Basis*. Contribution of Working Group I to the Third Assessment Report of the Intergovernmental Panel on Climate Change. HOUGHTON, J. T., DING, Y., GRIGGS, D. J., NOGUER, M., VAN DER LINDEN, P. J., DAI, X., MASKELL, K. & JOHNSON, C. A. (eds). Cambridge University Press, Cambridge.

IRVINE-FYNN, T. D. L., MOORMAN, B. J., SJOGREN, D. B., WALTER, F. S. A., WILLIS, I. C., HODSON, A. J., WILLIAMS, J. L. M. & MUMFORD, P. N. 2005. Cryological processes implied in Arctic pro-glacial stream sediment dynamics using principal components analysis and regression. *In:* HARRIS, C. & MURTON, J. B. (eds.): *Cryospheric Systems: Glaciers and*

Permafrost. Geological Society, London, Special Publications, **242**, 83–98.

JOHNSTON, G. H., LADANYI, B., MORGENSTERN, N. R. & PENNER, E. 1981. Engineering characteristics of frozen and thawing soils. *In*: JOHNSON, G. H. (ed.) *Permafrost: Engineering Design and Construction*. Wiley, New York, 73–147.

KAPLYANSKAYA, F. A. & TARNOGRADSKIY, V. D. 1986. Remnants of the Pleistocene ice sheets in the permafrost zone as an object for paleoglaciological research. *Polar Geography and Geology*, **10**, 257–266.

KIRKBRIDE, M. & DUGMORE, A. J. 2005. Late Holocene solifluction history reconstructed using tephrochronology. *In*: HARRIS, C. & MURTON, J. B. (eds.): *Cryospheric Systems: Glaciers and Permafrost*. Geological Society, London, Special Publications, **242**, 145–155.

KNIGHT, P. G. 1997. The basal ice layer of glaciers and ice sheets. *Quaternary Science Reviews*, **16**, 975–993.

KOZARSKI, S. 1959. O genezie chodzieskiej moreny czołowej (Summary: On the origin of the Chodzież end moraine). *Badania Fizjograficzne nad Polską Zachodnią*, **5**, 45–72. [English translation: On the origin of the Chodziez End Moraine. 1994. *In*: EVANS, D. J. A. (ed.) *Cold Climate Landforms*. Wiley, Chichester, 293–312.]

LACHENBRUCH, A. H. & MARSHALL, B. V. 1986. Changing climate: Geothermal evidence from permafrost in the Alaskan Arctic. *Science*, **234**, 689–696.

ŁOZIŃSKI, W. VON. 1909. Uber die mechanische Verwitterung der Sandsteine im gemassigten Klima. *Acad. Sci. Cracovie Bull. Internat. cl. sci. math et naturelles*, **1**, 1–25. [English translation: On the mechanical weathering of sandstones in temperate climates. *In*: EVANS, D. J. A. (ed.) *Cold Climate Landforms*. Wiley, Chichester, 119–134.]

ŁOZIŃSKI, W. VON. 1912. Die periglaziales fazies der mechanischen Verwitterung. *In*: *Comptes Rendus, XI Congrès Internationale Géologie*. Stockholm 1910, 1039–1053.

MACKAY, J. R. 1971. The origin of massive icy beds in permafrost, western Arctic coast, Canada. *Canadian Journal of Earth Sciences*, **8**, 397–422.

MACKAY, J. R. 1977. Pulsating pingos, Tuktoyaktuk Peninsula, N.W.T. *Canadian Journal of Earth Sciences*, **14**, 209–222.

MACKAY, J. R. 1989. Massive ice: Some field criteria for the identification of ice types. *In*: *Current Research, Part G, Geological Survey of Canada*, Paper 89-1G, 5–11.

MACKAY, J. R. 1998. Pingo growth and collapse, Tuktoyaktuk Peninsula area, western Arctic coast, Canada: A long-term field study. *Géographie physique et Quaternaire*, **52**, 271–323.

MACKAY, J. R. & DALLIMORE, S. R. 1992. Massive ice of the Tuktoyaktuk area, western Arctic coast, Canada. *Canadian Journal of Earth Sciences*, **29**, 1235–1249.

MACKAY, J. R., RAMPTON, V. N. & FYLES, J. G. 1972. Relic Pleistocene permafrost, western Arctic, Canada. *Science*, **176**, 1321–1323.

MARCHANT, D. R., LEWIS, A. R., PHILLIPS, W. M., MOORE, E. J., SOUCHEZ, R. A., DENTON, G. H., SUGDEN, D. E., POTTER JR, N. & LANDIS, G. P. 2002. Formation of patterned ground and sublimation till over Miocene glacier ice in Beacon Valley, southern Victoria Land, Antarctica. *Bulletin of the Geological Society of America*, **114**, 718–730.

MARSH, P. & WOO, M. K. 1981. Snowmelt, glacier melt and high Arctic streamflow regimes. *Canadian Journal of Earth Sciences*, **18**, 1380–1384.

MATHEWS, W. H. & MACKAY, J. R. 1960. Deformation of soils by glacier ice and the influence of pore pressure and permafrost. *Royal Society of Canada Transactions*, LIX, Series III, Section 4, 27–36.

MIALL, A. D. 1983. Glaciofluvial transport and deposition. *In*: EYLES, N. (ed.) *Glacial Geology*. Pergamon Press, Oxford, 168–183.

MELLOR, M. & TESTA, R. 1969. Effect of temperature on the creep of ice. *Journal of Glaciology*, **8**, 131–145.

MERCIER, D. & LAFFLY, D. 2005. Actual paraglacial progradation of the coastal zone in the Kongsfjorden area, western Spitsbergen (Svalbard). *In*: HARRIS, C. & MURTON, J. B. (eds.): *Cryospheric Systems: Glaciers and Permafrost*. Geological Society, London, Special Publications, **242**, 111–117.

MOORMAN, B. J. 2005. Glacier–permafrost hydrological interconnectivity: Stagnation Glacier, Bylot Island, Canada. *In*: HARRIS, C. & MURTON, J. B. (eds.): *Cryospheric Systems: Glaciers and Permafrost*. Geological Society, London, Special Publications, **242**, 63–74.

MOORMAN, B. J. & MICHEL, F. A. 2000. The burial of ice in the proglacial environment on Bylot Island, Arctic Canada. *Permafrost and Periglacial Processes*, **11**, 161–175.

MURTON, J. B. 2005. Ground-ice stratigraphy and formation at North Head, Tuktoyaktuk Coastlands, western Arctic Canada: A product of glacier–permafrost interactions. *Permafrost and Periglacial Processes*, **16** (in press).

MURTON, J. B. & KOLSTRUP, E. 2003. Ice-wedge casts as indicators of palaeotemperatures: Precise proxy or wishful thinking? *Progress in Physical Geography*, **27**, 155–170.

MURTON, J. B., WALLER, R. I., HART, J. K., WHITEMAN, C. A., POLLARD, W. H. & CLARK, I. D. 2005a. Stratigraphy and glaciotectonic structures of permafrost deformed beneath the northwest margin of the Laurentide Ice Sheet, Tuktoyaktuk Coastlands, Canada. *Journal of Glaciology*, **170** (in press).

MURTON, J. B., WHITEMAN, C. A., WALLER, R. I., POLLARD, W. H., CLARK, I. D. & DALLIMORE, S. R. 2005b. Basal ice facies and supraglacial melt-out till of the Laurentide Ice Sheet, Tuktoyaktuk Coastlands, western Arctic Canada. *Quaternary Science Reviews* (in press).

RAMPTON, V. N. 1974. The influence of ground ice and thermokarst upon the geomorphology of the Mackenzie-Beaufort region. *In*: FAHEY, B. D. & THOMPSON, R. D. (eds) *Research in Polar and*

Alpine Geomorphology. Proc. Third Guelph Symposium on Geomorphology, Geo-Abstracts, Norwich, 43–59.

RAMPTON, V. N. 1988. *Quaternary Geology of the Tuktoyaktuk Coastlands, Northwest Territories.* Geological Survey of Canada, Memoir 423.

RAMPTON, V. N. 1991. Observations on buried glacier ice and massive segregated ice, western Arctic coast, Canada: Discussion. *Permafrost and Periglacial Processes*, **2**, 163–165.

RAMPTON, V. N. & MACKAY, J. R. 1971. Massive ice and icy sediments throughout the Tuktoyaktuk Peninsula, Richards Island, and nearby areas, District of Mackenzie. Geological Survey of Canada, Paper 71–21.

SCHOTT, C. 1933. Die Formengestaltung der Eisrandlagen Norddeutschlands. *Zeitschrift fur Gletscher*, **21**, 54–98.

SHARPE, D. R. 1992. *Quaternary Geology of Wollaston Peninsula, Victoria Island, Northwest Territories.* Geological Survey of Canada Memoir 434.

ST-ONGE, D. A. & McMARTIN, I. 1999. The Bluenose Lake Moraine, a moraine with a glacier core. *Géographie physique et Quaternaire*, **53**, 287–295.

SVENDSEN, J. I., ALEXANDERSON, H., ASTAKHOV, V. I., ET AL. 2004. Late Quaternary ice sheet history of northern Eurasia. *Quaternary Science Reviews*, **23**, 1229–1271.

TSYTOVICH, N. A. 1975. *The Mechanics of Frozen Ground.* McGraw-Hill, New York.

VANDENBERGHE, J. 1983. Ice-wedge casts and involutions as permafrost indicators and their stratigraphic position in the Weichselian. *In*: *Permafrost: Fourth International Conference, Proceedings, Fairbanks, Alaska.* National Academy Press, Washington, DC, 1298–1302.

VIALOV, S. S. 1965. *The Strength and Creep of Frozen Soils and Calculations for Ice-Soil Retaining Structures.* Translation 76, US Army Cold Regions Research and Engineering Laboratory, Hanover, NH.

VINNIKOV, K. Y., ROBOCK, A., STOUFFER, R. J., WALSH, J. E., PARKINSON, C. L., CAVALIERI, D. J., MITCHELL, J. F. B., GARRETT, D., & ZAKHAROV, V. F. 1999. Global warming and Northern Hemisphere sea ice extent. *Science*, **286**, 1934–1937.

WALLER, R. I. 2001. The influence of basal processes on the dynamic behaviour of cold-based glaciers. *Quaternary International*, **86**, 117–128.

WALLER, R. I. & TUCKWELL, G. W. 2005. Glacier–permafrost interactions and glaciotectonic landform generation at the margin of the Leverett Glacier, West Greenland. *In*: HARRIS, C. & MURTON, J. B. (eds.): *Cryospheric Systems: Glaciers and Permafrost.* Geological Society, London, Special Publications, **242**, 39–50.

WASHBURN, A. L. 1979. *Geocryology: a Survey of Periglacial Process and Environments.* Edward Arnold, London.

WILLIAMS, P. J. & SMITH, M. W. 1989. *The Frozen Earth: Fundamentals of Geocryology.* Cambridge University Press, Cambridge.

WOO, M. K., MARSH, P. and POMEROY, J. W. 2000. Snow, frozen soils and permafrost hydrology. *Hydrological Processes*, **14**, 1591–1611.

YERSHOV, E. D. 1998. *General Geocryology.* Cambridge University Press, Cambridge.

Glacier–permafrost interaction in Arctic and alpine mountain environments with examples from southern Norway and Svalbard

BERND ETZELMÜLLER & JON OVE HAGEN

Department of Geosciences, Section of Physical Geography, University of Oslo, PO Box 1047, Blindern, N–0316 Oslo, Norway
(e-mail: bernde@geo.uio.no, joh@geo.uio.no)

Abstract: The interaction between glaciers and permafrost was long ago addressed for glaciers in Arctic regions. Analogies from modern environments have been used to understand landform development at the margins of Pleistocene ice sheets. During more recent decades many systematic measurements of permafrost in boreholes, geophysical soundings and temperature monitoring have revealed permafrost to be more abundant in many more high-mountain areas than previously thought. This suggests that permafrost may be a governing factor not only for periglacial landform evolution in these areas, but also, given the potential for glacier–permafrost interaction, for glacial landform generation. This paper presents and discusses observation and study results on the geomorphological significance of the interrelationship between glaciers and permafrost, in relation to geomorphological processes, landform generation and response of the system to climate fluctuations.

In the geo-scientific literature, processes and landforms related to glaciers and permafrost are often treated separately. However, both are part of the cryosphere, they often co-exist and therefore there is a potential for interaction. Thus, knowledge of these interactions promotes better understanding of glacial landform formation, and is crucial when discussing glacial and paraglacial processes and geotechnical hazards (Haeberli 1992). Interaction between glaciers and permafrost has been given only limited attention in scientific literature, even though such relationships have long been recognized, for example the role of permafrost in the development of push moraines in central Europe during the last glaciation (e.g., Gripp 1929) or more recently in the southern fringe of the Wisconsin ice sheet in North America (e.g., Bluemle & Clayton 1984; Clayton *et al.* 2001). Attention has focussed on the co-existence of glaciers in permafrost areas, mainly in the Arctic and/or former Pleistocene cryosphere environments. An exception to this is provided by the work of Liestøl (1977, 1996), who linked ground and glacier thermal regimes to catchment hydrology and pingo development on Svalbard, highlighting the importance of a fundamental understanding of glacier–permafrost interaction. During the last 20 years, it has been apparent that there is extensive permafrost in many high-mountain

environments, even at lower latitudes (e.g., Haeberli *et al.* 1993). Recently, boreholes revealed permafrost up to several hundreds of metres thick across a number of European mountain regions (Harris *et al.* 2003).

This paper presents observations from glaciers and permafrost in mountain environments of southern Norway and Svalbard. The paper focuses on mountainous environments, with cirque, valley or plateau glaciers, with terrestrial glacier tongues. The aim of the paper is to discuss field and laboratory results and certain well-known glaciological and geomorphological relationships conceptually, emphasizing the role of glacier–permafrost interaction, its relevance to selected geomorphological processes and their response to climate change. Furthermore, we want to draw together some general and hypothetical concepts, which may serve to stimulate further research within this field.

Permafrost definition

According to the International Permafrost Association (1998), permafrost is thermally defined as 'ground that remains at or below 0°C for at least two consecutive years'. Perennial ground ice or congelation ice is explicitly included in the permafrost definition, while glacier ice (sedimentary ice) is not (see also

From: HARRIS, C. & MURTON, J. B. (eds) 2005. *Cryospheric Systems: Glaciers and Permafrost.*
Geological Society, London, Special Publications, **242**, 11–27.
0305-8719/05/$15.00 © The Geological Society of London 2005.

Shumskiy 1964). From a purely thermal point of view at least, glacier ice at sub-pressure melting point temperatures would become a part of the permafrost environment. On the other hand, permafrost is not a precondition for the formation of sedimentary ice (in contrast to congelation ice), and thus, being part of the hydrosphere, glaciers are excluded. These definitions and their conceptual boundaries are of geomorphological importance and form the basis for the further discussion and hypotheses presented in this paper.

Glacier–permafrost interaction

Most glaciers present in permafrost regions are characterized by relatively low accumulation rates and a short ablation season, of 2–4 months, during summer. This leads to a smaller mass turnover and lower meltwater discharges compared with temperate regions (Bogen & Bønsnes 2003; Hagen *et al.* 2003). The lower activity is manifested in lower velocities, often resulting in smoother ice surfaces (low crevasse density), which often prevents linkage between supraglacial meltwater and englacial or subglacial drainage systems.

Most glaciers have areas with below-pressure-melting-point temperatures, and are thus connected to the permafrost environment. The distribution of cold ice in glaciers varies however, depending on the thickness of the ice body and the energy exchange processes due to phase transitions of water (Paterson 1994). In general, one can expect cold-based ice in glaciers when the equilibrium line altitude (ELA) is at or well below the lower limit of widespread permafrost in a region (mountain permafrost altitude, MPA; Fig. 1). Thin glaciers in Arctic or high-altitude regions with the MPA well below the glacier terminus are often entirely cold-based (Fig. 1a), while in warmer and wetter areas, where summer melting occurs, the ice in the accumulation area is warmed due to latent heat release by meltwater refreezing, resulting in less extensive (Fig. 1b, c) or even absent below-pressure-melting-point temperatures in the accumulation area (Fig. 1d; Paterson 1994). Glaciers where both temperate (temperatures at the pressure melting point) and cold (below-pressure-melting-point temperatures) ice is present are called *polythermal* glaciers. In such cases the marginal parts of the accumulation area

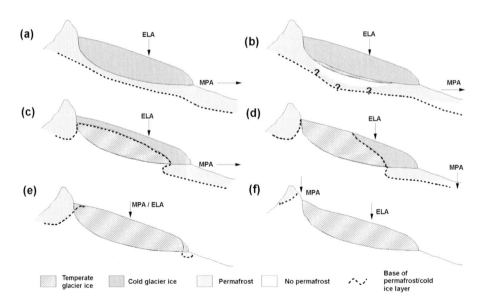

Fig. 1. Schematic illustration of thermal regimes in glaciers related to permafrost in mountain areas. (**a**) Entire cold glacier in a permafrost environment. MPA ≪ ELA. (**b**) Mostly cold-based glacier with temperate layers along the glacier-ground interface (e.g., Austre Brøggerbreen, Midtre Lovénbreen, Svalbard; Björnsson *et al.* 1996).
MPA ≪ ELA. (**c,d**) Polythermal glacier with temperate glacier ice along most of the bottom and in the accumulation area (e.g., Erikbreen, Finsterwalderbreen, Kongsvegen; Ødegård *et al.* 1992, 1997; Björnsson *et al.* 1996), MPA < ELA. This are the typical case described by Liestøl (1977) for most of the terrestrial glaciers on Svalbard. (**e**) Mostly warm-based glacier with some cold patches, either locally in the frontal areas (e.g., Midtdalsbreen, southern Norway) or in dry, cold accumulation areas (e.g., Alps). MPA ≈ ELA or MPA > ELA. For entirely temperate glaciers, MPA ≫ ELA.
(**f**) Åfotbreen, Nigardsbreen, southern Norway

may be frozen to the mountain side and head wall, and the terminal zone of the glacier may be frozen to the ground. Several examples of these glaciers exist in the Arctic, for instance on Svalbard (Björnsson *et al.* 1996) or in Arctic Canada (Blatter 1987). The size of the glacier area that is cold-based varies (Blatter & Hutter 1991). In sub-Arctic regions, there are reported to be glaciers with only very marginal cold areas, such as Midtdalsbreen (Hagen 1978; Liestøl & Sollid 1980) and Leirbreen (Harris & Bothamley 1984) in southern Norway. Where the ice is thin, such as along glacier margins slightly below the MPA, winter cold may penetrate the glacier ice. During the summer, this cold is not removed, leading to temperatures below the pressure melting point temperatures (Fig. 1e). Another thermal type occurs in high-altitude glaciers that start above and end below the MPA. These glaciers may have partly cold accumulation areas (Suter *et al.* 2001) if such areas are situated at altitudes with no or only very limited summer melting, such as for example in certain parts of the Alps. If the ELA is located well below the MPA, temperate glaciers dominate (Fig. 1f).

Glaciers have the ability to advect heat to the ice–bed interface in the accumulation areas, and thus can promote the development of closed or open taliks in the underlying sediments or bedrock (either totally enclosed by permafrost or connected with the unfrozen sub-permafrost drainage system). Liestøl (1977) (Fig. 2) defined this open talik as the characteristic case for Svalbard, where he described glacier melt-water penetration, subglacially into the ground, feeding sub-permafrost groundwater flow in the valleys, which locally form open-system pingos, pro-glacial winter icings and springs. In addition, thin ice covers and snow-fields or snow patches lying slightly below the MPA preserve subzero temperatures in the ground if they are sufficiently thin that the winter cold penetrates the snow/ice mass.

In summary, glaciers ending in a stable terrestrial permafrost environment will always be polythermal, as at least marginal areas of the glaciers will be cold-based.

Svalbard and southern Norway

Morphological visible interaction of glaciers and permafrost exists, especially in areas where glaciers end in a permafrost environment. As permafrost is thermally defined, climate forcing, the ability of the ground to conduct heat and the local geothermal gradient govern permafrost distribution and thickness.

On Svalbard, the mean annual air temperature (MAAT) at the coast is about $-5°C$, with low

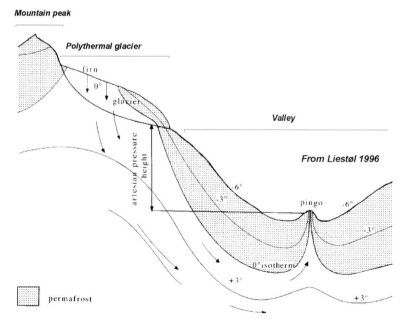

Fig. 2. This profile illustrates the relationship between glacier thermal regime and permafrost distribution in a typical valley setting on Svalbard. Reprinted from Open-system pingos in Spitsbergen by Liestøl from Norsk Geografisk Tidsskrift, wwwtandf.no/ngeog, 1996, 50, 81–84, by permission of Taylor and Francis AS.

winter temperatures and stable summer temperatures below 10°C. This results in a high number
of freezing degree-days (2000–3000°C days) in
relation to thawing degree-days (400–550°C
days; Humlum *et al.* 2003), and thus continuous
permafrost. Permafrost in the non-glacierized
central mountains of Svalbard reaches a thickness of more than 450 m, decreasing to below
150 m towards the coast (Liestøl 1977). This
means that all glaciers ending on land on
Svalbard extend into a permafrost environment.
The distribution of surficial material on Svalbard
is dominated by *in-situ* weathered material
(Kristiansen & Sollid 1987; Sollid & Sørbel
1988b) in the interior parts of the islands. Close
to the coast, sediments range from gravelly
beach sediments to finer-grained marine deposits, as a consequence of high post-Weichselian
marine limits (Mangerud *et al.* 1996). In the
marine sediments, saline pore water is common,
depressing the freezing point. This results in a
considerable amount of unfrozen pore water,
generating a relatively weak permafrost layer,
as observed in Svea (Gregersen *et al.* 1983;
Furuberg & Berggren 1988). The permafrost
thins towards the coast. There are no indications
of sub-sea permafrost. Recent investigations
indicate permafrost aggradation in response to
Holocene isostatic uplift (Humlum 2004). The
permafrost altitude that defines where permafrost
becomes abundant (lower regional MPA) lies
below or near sea level on Svalbard.

Permafrost in southern Norway is restricted to
high-mountainous areas. At a regional scale, the
MAAT of -3.5 to -4°C gives a good estimate
of the lower limit of regional, discontinuous
permafrost (King 1986; Ødegård *et al.* 1996;
Etzelmüller *et al.* 1998). However, snow cover
and other local climate effects make permafrost
abundant even in warmer or lower settings (e.g.
Sollid *et al.* 2003). Based on these boundary conditions, permafrost is mainly concentrated in a
50–100 km-wide zone following the main
mountain crest in a southwest-northeast direction
(Fig. 3). Only small areas east and west of this
zone have permafrost. The regional MPA
decreases from above 1600 m above sea level
(a.s.l.) in the west to below 1300 m a.s.l. in the
east, close to the Swedish border.

On the western side, several glaciers cover
high mountain areas, and the landscape is dominated by alpine relief types with high peaks and
deeply incised valleys. On the eastern side of
this zone, the area is dominated by mountain pla-
teaux situated below the lower regional MPA.
However, small mountain areas or single peaks
reach altitudes above this limit (Heggem *et al.*
2003). Patchy permafrost is abundant at much

lower altitudes than described above, e.g., in
palsa mires (Sollid & Sørbel 1998) or in coarse
slope deposits in extreme shadow (Liestøl 1965).
Surficial material in mountain areas is often thin
morainic or *in-situ*-weathered material. In
southern Norway, glaciers end both below and
above the MPA. In most maritime environments
with high mean annual precipitation and MAAT
of *c.* 0°C, at the glacier equilibrium line altitude
(ELA), temperate glaciers dominate, extending
below the lower regional MPA. In these areas
the ELA lies at or well below the MPA.
Glacier–permafrost interactions are most likely
when the ELA lies well above the MPA,
especially when the glacier snout terminates at
or above the MPA. In dry, cold areas, insufficient
precipitation is present for the formation of
glaciers. For southern Norway, therefore, a
west–east gradient is present with respect to
ELA, MPA, and glaciers within the permafrost
zone in the central areas (Fig. 3).

For both southern Norway and Svalbard there
is strong evidence that the present glacier coverage is a response to late Holocene cooling, which
started about 4000–5000 year BP. In southern
Norway, it has been shown that there were few
glaciers during the Holocene climatic optimum
(Dahl & Nesje 1994; Nesje *et al.* 2000), while
on Svalbard glacier coverage was certainly considerably less, and many smaller glaciers even
disappeared, with permafrost-free areas along
the coast and in low-lying valley bottoms. A
comprehensive review of Holocene climate
change in the northern polar and sub-polar hemisphere is given in Humlum *et al.* (2005) and
Humlum (2005).

Geomorphological processes: general observations

General landform type and distribution pattern

Svalbard. For most glaciers ending on land on
Svalbard, one or occasionally two prominent
arcs of end moraines, often up to 50 m in
height, enclose the glacier foreland, indicating
the maximum Holocene extension of these glaciers (Figs 4a and 5a). The few dates available
for Svalbard indicate that the outermost postglacial position was reached a couple of times
since 2500 years BP, and culminated during the
Little Ice Age (LIA) at the turn of the twentieth
century (Furrer 1992; Snyder *et al.* 2000;
Humlum *et al.* 2005). Between these outermost
ridges and the present glacier front, areas with
minor glacial mounds, flow tills and outwash

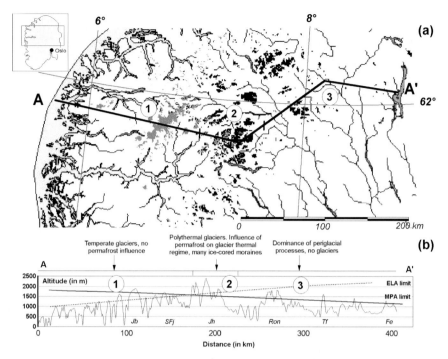

Fig. 3. (**a**) The distribution of alpine permafrost in southern Norway. Probable permafrost areas are displayed in dark shading, recent glacier coverage in grey shading. The solid line shows the profile line displayed in (b). The method of deriving permafrost areas used for southern Norway is described in Etzelmüller *et al.* (1998, 2003). The inset map denotes the position of the drawing in southern Norway. The numbers correspond to those described in (b). (**b**) Profile through southern Norway including the lower limit of alpine permafrost and the limits of the ELA. A spatial distribution of the ELA in southern Norway was generated based on an ELA contour map published in Liestøl (1994). The profile is interpreted to distinguish three major morphogenetic zones in relation to glacial and periglacial processes (zones 1–3). *Jb* = Jostedalsbreen glacier; *SFj* = Sognefjellet mountains; *Jh* = Jotunheimen; *Ron* = Rondane mountains. Based on Etzelmüller *et al.* (2003), slightly changed

plains are found. Both the end moraines and much of the flow till areas are cored or underlain by glacier ice (Boulton 1967; Sollid & Sørbel 1988a; Etzelmüller *et al.* 1996; Lyså & Lønne 2001). Fluting and stream-lined ridges in this marginal zone indicate temperate conditions during maximal glacial extension (Glasser & Hambrey 2001). Multi-temporal altitude comparison by air photos for several glacier forelands on Svalbard demonstrated a surface lowering along erosive river channels and along ice-cliffs, displaying the stability of areas with debris-covered ice vs melting of areas with exposed ice over periods of several decades (Etzelmüller 2000). When deposited on steep slopes, some ice-cored terminal or lateral moraines evolve into creeping bodies, commonly termed glacier-derived rock glaciers (Fig. 5b). These forms are abundant in the drier central areas of the Spitsbergen Island (Kristiansen & Sollid 1987),

and their dynamics are often decoupled from the glacier dynamics.

In some Svalbard glaciers, push moraines (Gripp 1929; Stauchendmoränen) or composite ridges (Benn & Evans 1998) are observed outside the main ice-cored moraines. These are normally less than 10 m in height, and occur in distinct zones at the limit of the glacier foreland (Figs 4a and 5c). The forms may be defined as pro-glacial sediments, pushed into ridges by advance of the glacier margin, and showing thrust structures (Hambrey *et al.* 1996) or other deformation structures (Boulton 1986; Etzelmüller *et al.* 1996). These moraines are, however, not in direct contact with the glacier snout today. Numerous sedimentological studies show folding and thrusting in both coarse-grained and fine-grained pro-glacial sediments on Svalbard (Fig. 5d; Croot 1988; Hart & Watts 1997; Boulton *et al.* 1999).

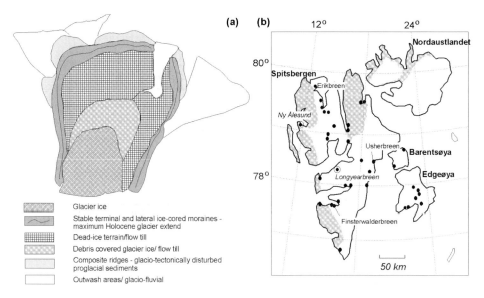

Fig. 4. (**a**) Schematic draft of a typical terrestrial glacial marginal area on Svalbard. The pro-glacial areas contain one or two prominent arcs of ice-cored moraines and a dead-ice zone with flow tills. An outermost zone of composite ridges is identified on numerous glaciers; however, they are mostly below the marine limit and preferably in areas dominated by sedimentary bedrock. (**b**) Location of composite ridges in the sense of large-scale deformed and thrusted pro-glacial sediments on Svalbard. The shaded area denotes pre-Devonian basement rocks, while the other areas are dominated by sedimentary bedrock. Based on Etzelmüller *et al.* (1996), slightly changed

Both ice-cored and pro-glacial push moraines co-exist in front of numerous Svalbard glaciers (Fig. 4a). The pro-glacial push moraines on, for example, Erikbreen display relict forms, generated during former advances. On Usherbreen, active formation of push moraines was observed and measured during a glacier surge in 1985 (Hagen 1988). O. Liestøl (personal communication) recognized at an early stage that these types of moraine on Svalbard are restricted to areas below the Holocene marine limit. They also seem more abundant in areas dominated by sedimentary rocks. Ice-cored moraines, however, develop practically in all glaciers margins ending on land on Svalbard.

Southern Norway. In southern Norway, stable ice-cored moraines are abundant in the Jotunheimen and Dovrefjell areas (Østrem 1964). These ice-cored moraines are connected mostly to minor cirque glaciers. These glaciers end within environments where regional permafrost modelling confirmed by site investigations suggests widespread permafrost (Ødegård *et al.* 1996; Etzelmüller *et al.* 1998; Isaksen *et al.* 2002). Composite ridges in the sense described in Svalbard have not been observed in southern Norway. Haeberli (1979), however, indicated that moraine features in front of many glaciers in the high mountains of Scandinavia might

partly be a result of sediment deformation in a permafrost environment.

The glacial particulate sediment flux system

Most measurements of fluvial sediment flux have been carried out in relation to temperate glaciers within a non-permafrost environment (Lawson 1993). More recently, however, flux from poly-thermal glacial settings in Svalbard (Bogen & Bønsnes 2003) has been measured. In general, these studies indicate that, in temperate glacier systems, most of the sediments produced by the glacier are evacuated by meltwater, and sediment flux during summer decreases, despite increasing discharge (Liestøl 1967; Østrem 1975). In permafrost areas with cold or polythermal glaciers, a closer relationship has been observed between discharge and sediment concentrations throughout the melting season (e.g. Vatne *et al.* 1995; Bogen 1996). This observation indicates the more limited ability of the meltwater to exhaust subglacially stored material, and the general availability of sediments for fluvial erosion beneath and in front of these glaciers (Vatne *et al.* 1995; Hodgkins 1997; Hodson *et al.* 1997). Despite the more limited glacier activity, the shorter meltwater season and the lower meltwater production, overall erosion

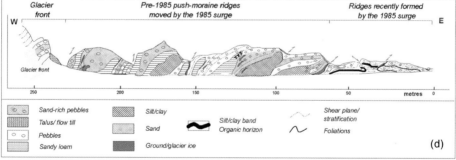

Fig. 5. (**a**) Moraine area in front of Marthabreen in the Reindalen valley on central Spitsbergen, south of Longyearbyen. Most of the glacier fore-field is underlain by glacier ice, and the outer-most ridges of the area are dominated by ice-cored moraines (photograph: Jon Ove Hagen, 1986, 1994). (**b**) Marginal zone of a unnamed glacier, Reindalen (photograph Leif Sørbel, 1994). The moraine area is ice-cored, creeping down-slope independently of the glacier dynamics, forming a rock glacier. (**c**) Glacier marginal zone of Usherbreen, eastern Spitsbergen. The ridges are pushed and moved by a surge advance of the glacier (photo: Jon Ove Hagen, 1985). The white line indicates the location of the profile shown in (d). (**d**) Section logged in the moraine area of Usherbreen. The structure of the ridge system could be observed in a cross section carved through the ridge system by a water channel. The sediment layers are highly dipped and deformed, folded ice layers are frequently observed (based on Etzelmüller *et al.* 1996, slightly changed)

rates from poly-thermal glacier catchments on Svalbard are remarkably high, with values up to 1 mm per annum (Sollid *et al.* 1994; Hallet *et al.* 1996; Hodgkins *et al.* 2003), even for small, relatively inactive and presumably cold glacier catchments like Longyearbreen (Etzelmüller *et al.* 2000). These high values are mainly attributed to the easy availability of sediments in the dead-ice and flow till-dominated glacier fore fields.

Geomorphological processes: conceptual models and hypotheses

The distribution of sub pressure-melting point temperatures in glaciers is recognized as a crucial factor for the production, incorporation, transport and sedimentation of glacial material (Weertman 1961; Boulton 1972). This factor has also recently been identified in sedimentological work on small terrestrial Svalbard glaciers (Hambrey et al. 1999; Glasser & Hambrey 2001). In the following section, some of these topics are outlined in the framework of glacier–permafrost interaction.

Ice-cored moraines

At the glacier bed the transition zone between temperate and cold ice favours the adfreezing of sediments into basal ice and their transport to englacial and ultimately supraglacial locations. At this thermal transition, the incorporation of sediments due to thrusting and along shear planes is also facilitated by high compressive flow, resulting in a frontal ice-compression zone in which shear zones and folding of ice develops (Hambrey et al. 1999; Glasser & Hambrey 2001). This leads to en-glacial transport of sediments towards the ice surface. These processes cause the accumulation of often coarse-grained or even stratified debris in and on the glacier snout. In permafrost environments the thickness of accumulated supraglacial debris may exceed the climate-controlled active-layer thickness, and thus protect the ice below from melting, resulting in the development of ice-cored moraines. This pattern is commonly described from polythermal glaciers in the high-Arctic (Johnson 1971; Souchez 1971; Sollid & Sørbel 1988a; Etzelmüller et al. 1996; Lyså & Lønne 2001; Sletten et al. 2001). In southern Norway, comparable processes and landform developments are observed on glaciers in the central high-mountain area described earlier (Østrem 1964), where permafrost is widespread.

The survival time of these ice-cored moraines is related to the thickness of the sediment cover. If the temperature under the material cover is below 0°C throughout the year, the ice will not melt, and the ice-core will be stable. In temperate glacier marginal settings, ice-cored moraines are also observed, especially in front of surging glaciers, in areas of high material production and sedimentation and in areas of meltwater flooding events that cover the ice with sediments, e.g., on Iceland (Krüger & Kjær 2000; Kjær & Krüger 2001). However, these ice-cored structures are prone to rapid mass wasting and are not stable,

decaying over decades (Krüger & Kjær 2000). Ice-cored moraines in a permafrost setting are, therefore, prominent features and stable over long time periods. Ice-core decay may first happen during a climatic warming and with resulting increase in of the active-layer thickness. Thus, stable ice-cored moraines are restricted to climatically stable areas and are thus good permafrost indicators.

Close to the fringe of the lower limit of mountain permafrost, adfreezing of material is observed in very marginal areas of glaciers, building asymmetric moraine morphology, often in the form of annual moraines. An example is Midtdalsbreen in the Finse area of southern Norway, a temperate outlet glacier of the Hardangerjökulen ice cap. Here, thermistor measurements have shown that the lower-most, up to 20 m thick area of the glacier front is cold-based (Hagen 1978; Liestøl & Sollid 1980). Annual moraines built up since the 1960s show clear melt-out of frozen subglacial layers (Andersen & Sollid 1971). Harris & Bothamley (1984) and Matthews et al. (1995) describe similar processes in front of Leirbreen and Styggedalsbreen, western Jotunheimen, respectively. These examples discussed above illustrate that the ice-marginal thermal regime is critically important in controlling processes of sedimentation.

Proglacial folding and thrusting

During glacier advances, compressive stresses can be transmitted by the glacier into pro-glacial sediments that may locally result in large-scale deformation and thrusting (Hagen 1988; Bennett 2001). The scale of this process depends on the sub-surface geo-technical conditions (sediment strength and structure), and the magnitude and duration of applied stress. Sediment deformation will occur if glacier-induced stress exceeds the sediment creep strength. The creep strength of the frozen sediment, in a permafrost environment, is highly dependent on liquid water content, ice content and the duration of applied stress (Tsytovich 1975; Williams & Smith 1989). The ratio of liquid water to solid ice within is in turn dependent on sediment type, and fine-grained sediment will contain more liquid water than coarse-grained sediment at the same temperature. In exposed marine sediments, the unfrozen water content also depends on salt content in the pore water, which depresses the freezing point (Tsytovich 1975). Cohesion is the dominant factor controlling the shear strength at sub-zero temperatures. Coarse-grained sediments have no cohesion in the unfrozen state, their strength characteristics depending on effective normal strass and the internal angle of friction. Once frozen, ice-bonds between grains provide considerable cohesion, and sediment strength increases. In a situation with a high frozen pore water content, contacts between grains can be partly or wholly replaced by ice-bonds (Tsytovich 1975). Frictional strength is then reduced and the creep characteristics of the soil increasingly determine the sediment strength. As ice has low long-term shear strength, layers of supersaturated coarse-grained sediments, e.g., outwash gravels, may then deform plastically under low applied stress. In the case of unfrozen fine-grained cohesive sediments, frictional shear strength is often lower but cohesion higher. In their unfrozen state, such soils are weaker than coarser sediments, and their low permeability allows pore water pressure to rise when compressive stresses area applied. Once frozen, ice-bonds increase the cohesion and sediment strength increases. The sediment then acts more like a rigid mass, allowing the transfer of stresses over longer distances.

In Svalbard, the deformation of frozen proglacial sediments often causes formation of large-scale composite ridges, showing evidence of internal deformation or block-thrusting over large areas across the front of terrestrial glacier snouts (Hagen 1988; Hambrey *et al.* 1996; Hart & Watts 1997; Boulton *et al.* 1999). Such moraines are particularly abundant in fine-grained saline coastal sediments below the marine limit (Fig. 4b). In these areas the strength of the partially frozen sediments is lower than that of the deforming glacier ice. The major décollement plane is often within sub-permafrost unfrozen sediments where porewater pressures are raised and effective stresses therefore are low. This is possible at glaciers near the coast on Svalbard, where permafrost is probably thin and in some cases is still aggrading. Humlum (2004) presents evidence that the permafrost in the major valleys started to aggrade at the present marine limit about 3500–3000 years BP in response to the general Holocene cooling in the Arctic. In the light of this, most of the composite ridges on near-coast Svalbard glaciers were probably developed earlier in the present phase of permafrost formation, when coastal permafrost was thinner than today.

Conceptual relationships between climate, thermal regime and landform generation

The relationship between climate, glacier thermal regime and glacial landform generation is conceptually exemplified in Fig. 6a. Based on this concept, zones of glacier processes in relation to temperature (presence or absence of permafrost) are outlined (Fig. 6b). A distinction is made between entirely cold glaciers, polythermal glaciers with cold glacier margins, glaciers with patchy zones of marginal adfreezing and glaciers dominated by marginal basal melting. The likely occurrence of a zone of composite ridges in the glacier foreland is suggested to decrease on both ends of the temperature range. If temperatures are low, permafrost becomes thick and stiff, decreasing its ability to deform or thrust the proglacial zone. The deeply frozen ice-cemented sediments have higher shear strengths than the glacier ice and are unlikely to be deformed by it. At the warmer extreme of the thermal continuum permafrost becomes sporadic and large-scale thrusting in the sense described in this paper becomes less likely. Thus, both ice-cored moraines and composite ridges can be recognized as indicators of permafrost conditions during their formation.

Influence of permafrost on the particulate sediment transfer system

In permafrost environments a thick material cover may preserve underlying ice over long time periods. Together with high extra-glacial material production and relatively low meltwater flux, permafrost favours the storage of sediments in the glacier forefields. However, water may locally remove the loose cover material, or mass movement processes may expose ice-cores, accelerating ice-core decay under permafrost conditions, and resulting in the formation of hummocky terrain (e.g. Sollid & Sørbel 1988a). These processes result in a chaotic dead ice terrain, with flow-tills and re-sedimentation processes (Boulton 1968; Etzelmüller 2000; Sletten *et al.* 2001). Thus, in such environments much material is mobilized, and available for erosion. Permafrost and ice-cored glacier marginal areas therefore form an important regulator of the sediment transfer system in which the particulate sediment evacuation is less likely to suffer sediment exhaustion during the summer. Material release depends on the removal of material protecting the ice-core, and this is a process happening through the summer melting season independent of glacial activity. Thus, even with low erosional potential, cirque and valley glacier catchments in permafrost environments have relatively high material transport and long-term erosion rates as measured by meltwater evacuation of solids (e.g. Hallet *et al.* 1996).

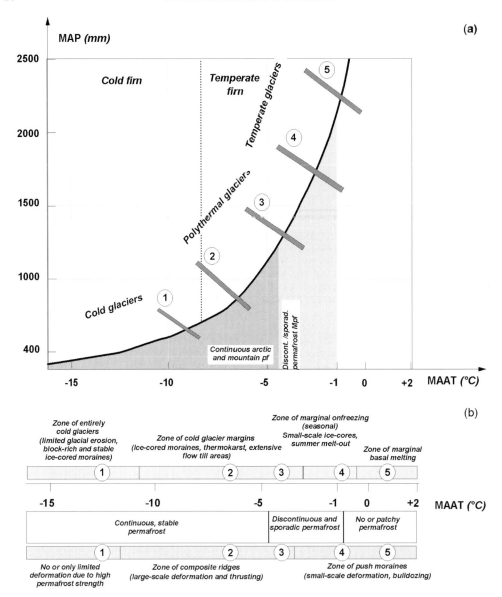

Fig. 6. (**a**) Schematic illustration of the relationship between MAAT and mean annual precipitation (MAP) at the ELA (solid line), and between (mountain) permafrost and thermal regime of the glaciers (based on Barsch 1996, modified). Above the solid line, glaciers dominate; while below the line permafrost dominates in colder environments due to lower precipitation. The permafrost-dominated regions are delineated into continuous and discontinuous/sporadic. The approximate division was set to MAAT < -4°C and < -1°C, respectively. These values may be widely valid in a regional context; however, locally large variations are widely reported due to the varying influence of topographic expositions, surficial material and snow distribution. The rectangles illustrate possible glacier extensions within this diagram. 1, Cold glaciers in continuous permafrost; 2, polythermal glaciers in continuous permafrost (e.g., on Svalbard); 3, polythermal glaciers advancing in discontinuous permafrost (e.g., high mountain areas of Scandinavia or the Alps); 4, temperate glaciers advancing through the discontinuous mountain permafrost zone; marginal permafrost at the front can be observed due to favourable local climatic conditions (topography, snow drift etc. e.g., Midtdalsbreen, Norway); 5, temperate glacier (e.g., Nigardsbreen). (**b**) Simplified illustration of the relationship between permafrost zones (indicated with MAAT) and marginal glacial–geomorphological processes. The numbers correspond to the glacier positions in (a). Pf = Permafrost

Response of the geomorphological process system to climate change

Both glaciers and permafrost respond to changing thermal boundary conditions that are largely climatically controlled. Glaciers react particularly to changing summer temperature and winter precipitation. Permafrost, being thermally defined, is controlled by the complex energy transfer processes at the boundary between the atmosphere and the earth surface. A geomorphologically important point is that glaciers show a relatively rapid response to climate change (years to tens of years), depending on glacier size and geometry. Permafrost, on the other hand, shows a comparatively slow response (hundreds to thousands of years), due to the dampening of diurnal to annual ground surface temperature variation with depth, which is controlled mainly by snow cover, the water content and the thermal properties of the surface sediments (Lachenbruch & Marshall 1986). This means, following the relationship illustrated in Fig. 1, that the shift of the ELA is normally faster than the shift of the MPA. This leads in turn to a potentially complex system of relationships between the thermal regimes of glaciers and permafrost.

A glacier advance caused by summer cooling or increased winter precipitation over proglacial permafrost leads to a warming of the ground over which the glacier has advanced. Advancing glaciers will introduce a more effective thermal buffer zone between atmosphere and lithosphere, and thus have both an insulating and a warming effect due to the potential advective heat transport of ice masses. This often leads to a warming of the overrun sub-glacial permafrost system despite cooling of the atmosphere. This was probably the case for the advance of glaciers during, for example, the LIA in many mountain and Arctic environments (Glasser & Hambrey 2001; Guglielmin 2004). During the cooler LIA period permafrost was more extensive in Norway, as a result of a lower mean annual ground surface temperature (MAGST). An empirical modelling approach based on distributed air temperatures indicated that the area affected by permafrost in southern Norway would have almost doubled with a 1 °C decrease of air temperature (Etzelmüller et al. 2003). However, since the LIA, climate cooling has also been associated with glacier advance, and the actual area of permafrost has increased by less than this amount. However, as an advance of an glacier over permafrost results first in a decrease in the ground temperature gradient, meaning that the warming is extremely slow, permafrost can survive for a very long time even under a temperate glacier.

Glacier retreat led to a cooling of the ground thermal regime in the newly exposed proglacial zone, and thus permafrost aggradation. In addition, initially temperate glaciers retreating into the marginal permafrost zone develop cold glacier snouts (Björnsson et al. 1996; Glasser & Hambrey 2001). Thus, in many areas, permafrost may have formed in the glacier marginal zones during glacier retreat, as recently documented by Kneisel (2003). In high mountain areas of southern Norway, where glaciers have been retreating into the permafrost zone since the LIA, this has been the case. Morphologically, this transition from temperate to cold-margin glaciers is evidenced by glacial landforms indicative of temperate glacier beds (e.g., flutings or striae) in areas which at present are underlain by permafrost (e.g. in Jotunheimen; Erikstad & Sollid 1990). Similar observations have also been reported from Svalbard by, for example, Glasser and Hambrey (2001), highlighting the consequences of the thermal regime change on the style of material entrainment and sedimentation.

Thinning of polythermal glaciers may result in their becoming entirely cold-based. Examples of this have been reported from Svalbard (Björnsson et al. 1996). A general thinning of the glaciers may therefore lead to a partial or total re-freezing of taliks below the glaciers and thus a restriction in potential sub-permafrost ground water inflow from glaciers. Subglacial erosion then stops in former temperate areas, with an associated increase in the potential for adfreezing of subglacial sediments (Weertman 1961).

As described earlier in this paper, glacial sediments in the permafrost zone are stored in extensive dead-ice terrain and ice-cored moraines over long time periods (decades to centuries), under steady climatic conditions and without erosive disturbance of the surface. During climate warming, a thicker active layer and thus an enhanced differential melting of debris-covered ice can be expected. Therefore, erosion of formerly permafrost-bonded sediment can be expected during warmer climate. These relationships are shown conceptually in Fig. 7.

Climate warming is also generally associated with reduction in glacier size. During such a phase glacial erosion rates will be reduced and meltwater discharge will increase, at least periodically, due to higher ice melting. Owing to lower sub-glacial material production, particulate suspended sediment concentrations in melt water rivers will decrease. This relationship has widely been used in analyses of lacustrine

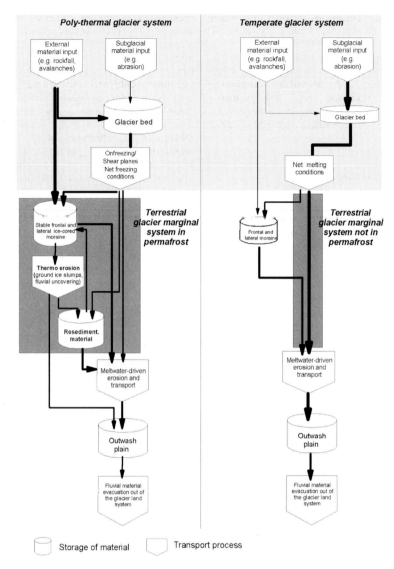

Fig. 7. Idealized schematic flow scheme of particulate material transport in cold/polythermal and temperate glacier environments. The grey shaded area denotes the glacier body, while the non-shaded areas display the terrestrial pro-glacial environment with and without permafrost. The thickness of the lines symbolizes the relative amount of material transported. The flow scheme illustrates that in permafrost environments a relatively larger portion of available material is routed through the ice-cored moraine system, while in the non-permafrost area most of the material is evacuated through the glacio-fluvial system. This affects the relationship between glacier activity and fluvial evacuation of particulate sediments out of glacierized catchments. During climate warming the two systems respond differently. Based on Etzelmüller et al. (2000), modified

sediment cores with respect to Holocene climate change (Nesje et al. 2001). However, in the permafrost zone, this reduced direct glacier sediment supply may well be compensated for by more rapid release of the sediments stored in the dead ice marginal zones and ice-cored moraines. On some glaciers on Svalbard, for instance, disintegration of ice-cored moraines has increased in the last 30 years, and the amount of material potentially mobilized by these processes nearly equals the measured suspended sediment yield from the investigated catchments (Etzelmüller 2000). Hodgkins et al. (2003) recently demonstrated the absence of a straightforward relationship between melt water production sediment yield from high-Arctic catchments.

In the context of paraglacial activity in high-alpine environments (Church & Ryder 1972; Ballantyne 2002), permafrost reduces the variability of sediment yield and promotes sediment storage in over longer time periods. Sediment stores are mobilized during permafrost degradation.

Concluding remarks

The distribution of sub-zero temperatures within glacier bodies and in the surrounding landscape is crucial for understanding geomorphological processes with respect to sediment production, transport and deposition in modern arctic and high-alpine glacierized environments. Within the system differing response times to climate change result in a complex ground and glacial thermal regime, influencing landform generation and preservation. There is a close interaction between the glacial and the periglacial system, affecting all chains of the sediment cascade system within landform evolution. In the framework of understanding geomorphological processes and response of the system to climate change, glaciers and permafrost systems must be regarded as an integrated part of the analyses (Fig. 8).

The proposed influence of permafrost on the sediment transfer system also provides constraints on lacustrine sedimentation. In studies on lake sediments, the signal in the lake cores indicates good correlation between glacier presence, size and the sediment record (Nesje et al. 2000). The glaciers referred to in these studies are temperate, and the lake cores allowed a detailed reconstruction of Holocene glaciers and, thus, climate history. Catchments containing cold and polythermal glaciers may also give higher sediment yields signal than those with no glacier at all (Svendsen et al. 1989; Barsch et al. 1994; Svendsen & Mangerud 1997; Bogen & Bønsnes 2003); however, the relationships between glacier variations, thermal regime and particulate sediment flux from such glacierized basins are likely to be complex and remain to be evaluated in more detail.

The process interrelationships discussed in this paper are crucial for understanding both the modern and former landscape formation in Scandinavia. The recent decades have provided an increased recognition of cold-based ice in both interior and marginal parts of the Scandinavian ice sheets (Sollid & Sørbel 1984, 1994; Kleman 1994), explaining landform preservation, pro-glacial sediment accumulation and deformation. In southern Norway at present, permafrost persist during glaciations or extends over at least twice the area of the glacierized zone. During considerable time spans in the Pleistocene it can be expected that permafrost affected large areas, including the mountainous zones, since it is likely that permafrost developed during glacier build-up, and that permafrost persist during glaciations or again formed during the retreat of the Pleistocene ice sheets. Therefore, it is likely that permafrost–glacier interaction influenced landform development over considerable time spans during the Pleistocene, both at microscales (moraine formation, slope processes) and macroscales (valley formation, strandflats).

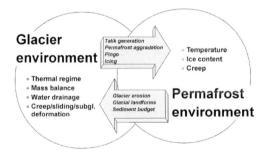

Fig. 8. Simplified diagram illustrating the coupling of cryospheric systems. Glaciers influence on the thermal regime of permafrost areas, resulting in permafrost degradation (talik) or aggradation and specific periglacial landforms such as open-system pingos. Permafrost influences glacial–geomorphological processes in terms of reduced subglacial erosion, specific glacial landform pattern and sediment budgets

This paper is based on a presentation given at the symposium on cryogenic systems held at the Geological Society cryospheric systems. London, in January 2003 by the first author. Most of the examples from Svalbard and southern Norway are based on field studies and observations, partly published in main co-authorships with J. L. Sollid, R. Ødegård, G. Vatne, K. Melvold and I. Berthling. The late Professor Olav Liestøl from the former Department of Physical Geography, University of Oslo, early realized the importance of permafrost on glacial geomorphological processes, and his work as glaciologist at the Norwegian Polar Institute (NPI) and teacher at the former Department of Physical Geography is a fundamental basis for this paper. Professor Emeritus J. L. Sollid, University of Oslo, Norway, over many years focused on the importance of glacial thermal regime in landform generation and deglaciation studies, and initiated the study of mountain permafrost in Norway at the former Department of Physical Geography, University of Oslo. N. Glasser and B. Rea reviewed this paper, and their numerous comments improved it considerably. O. Humlum gave valuable comments and new input to a later revised version of the paper, which the authors highly appreciate. We want to thank all mentioned persons and institutions.

References

ANDERSEN, J. L. & SOLLID, J. L. 1971. Glacial chronology and glacial geomorphology in the marginal zones of the glaciers Midtdalsbreen and Nigardsbreen, Southern Norway. *Norsk geografisk Tidsskrift*, **25**, 1–38.

BALLANTYNE, C. K. 2002. Paraglacial geomorphology. *Quaternary Science Reviews*, **21**, 1935–2017.

BARSCH, D. 1996. *Rock glaciers*, Springer, Berlin.

BARSCH, D., GUDE, M., MÄUSBACHER, R, SCHUKRAFT, G. & SCHULTE, A. 1994. Recent fluvial sediment budgets in glacial and periglacial environments, NW Spitsbergen. *Zeitschrift für Geomorphologie*, **97**(suppl.), 111–122.

BENN, D. I. & EVANS, D. J. A. 1998. *Glaciers and Glaciations*, Arnold, London.

BENNETT, M. 2001. The morphology, structural evolution and significance of push moraines. *Earth Science Reviews*, **53**(3–4), 197–236.

BJÖRNSSON, H., GJESSING, Y., HAMRAN, S.-E., HAGEN, J. O., LIESTØL, O., PALSSON, F. & ERLINGSSON, B. 1996. The thermal regime of sub-polar glaciers mapped by multi-frequency radio-echo sounding. *Journal of Glaciology*, **42**(140), 23–32.

BLATTER, H. 1987. On the thermal regime of an Arctic valley glacier: A study of White Glacier, Axel Heiberg Island, N.W.T., Canada. *Journal of Glaciology*, **33**(114), 200–211.

BLATTER, H. & HUTTER, K. 1991. Polythermal conditions in Arctic glaciers. *Journal of Glaciology*, **37**(126), 261–269.

BLUEMLE, J. P. & CLAYTON, L. 1984. Large-scale glacial thrusting and related processes in North Dakota. *Boreas*, **13**, 279–299.

BOGEN, J. 1996. Erosion rates and sediment yields of glaciers. *Annals of Glaciology*, **22**, 48–52.

BOGEN, J. & BØNSNES, T. E. 2003. Erosion and sediment transport in High Arctic rivers, Svalbard. *Polar Research*, **22**(2), 175–189.

BOULTON, G. S. 1967. The development of a complex supraglacial moraine of the margin of Sørbreen, Ny Friesland, Vestspitsbergen. *Journal of Glaciology*, **6**, 717–735.

BOULTON, G. S. 1968. Flow tills and related deposits on some Vestspitsbergen glaciers. *Journal of Glaciology*, **7**, 391–412.

BOULTON, G. S. 1972. *The Role of Thermal Regime in Glacial Sedimentation*. Institute of British Geography Special Publication no 4, 1–19.

BOULTON, G. S. 1986. Push moraines and glacier contact fans in marine and terrestrial environments. *Sedimentology*, **33**, 677–698.

BOULTON, G. S., VAN DER MEER, J. J. M., BEETS, D. J., HART, J. K. & RUEGG, G. H. J. 1999. The sedimentary and structural evolution of a recent push

moraine complex: Holmstrømbreen, Spitsbergen. *Quaternary Science Reviews*, **18**, 339–371.

CHURCH, M. & RYDER, J. M. 1972. Paraglacial sedimentation: A consideration of fluvial processes conditioned by glaciation. *Geological Society American Bulletin*, **83**, 3059–3071.

CLAYTON, L., ATTIG, J. W. & MICKELSON, D. M. 2001. Effects of late Pleistocene permafrost on the landscape of Wisconsin, USA. *Boreas*, **30**, 173–188.

CROOT, D. G. 1988. Glaciotectonics and surging glaciers: A correlation based on Vestspitsbergen, Svalbard, Norway. *In*: CROOT, D. G. (ed.), *Glaciotectonics: Forms and Processes*, Balkema, Rotterdam, 49–62.

DAHL, S. O. & NESJE, A. 1994. Holocene glacier fluctuations at Hardangerjøkulen, central southern Norway: A high-resolution composite chronology from lacustrine and terrestrial deposits. *The Holocene*, **4**, 269–277.

ERIKSTAD, L. & SOLLID, J. L. 1990. *Memurubreene, Jotunheimen – glasialgeomorfologi 1:6500*, Norsk Institutt for Naturforskning/Geografisk institutt, UiO, Oslo.

ETZELMÜLLER, B. 2000. Quantification of thermo-erosion in pro-glacial areas – examples from Spitsbergen. *Zeitschrift für Geomorphologie*, **44**(3), 343–361.

ETZELMÜLLER, B., HAGEN, J. O., VATNE, G., ØDEGÅRD, R. S. & SOLLID, J. L. 1996. Glacier debris accumulation and sediment deformation influenced by permafrost, examples from Svalbard. *Annals of Glaciology*, **22**, 53–62.

ETZELMÜLLER, B., BERTHLING, I. & SOLLID, J. L. 1998. The distribution of permafrost in Southern Norway; a GIS approach. *In*: LEWKOWICZ, A. G. & ALLARD, M. (eds), *Seventh International Conference on Permafrost, Proc.*, Collection Nordicana, Centre d'Etudes Nordiques, Universite Laval, Quebec, 251–257.

ETZELMÜLLER, B., ØDEGÅRD, R. S., VATNE, G., MYSTERUD, R. S., TONNING, T. & SOLLID, J. L. 2000. Glacier characteristic and sediment transfer systems of Longyearbyen and Larsbreen, western Spitsbergen. *Norsk geografisk Tidsskrift*, **54**(4), 157–168.

ETZELMÜLLER, B., BERTHLING, I. & SOLLID, J. L. 2003. Aspects and Concepts on the Geomorphological Significance of Holocene Permafrost in Southern Norway. *Geomorphology*, **52**(1–2), 87–104.

FURRER, G. 1992. Zur Gletschergeschichte des Liefdefjords. *Stuttgarter Geographische Studien*, **117**, 267–278.

FURUBERG, T. & BERGGREN, A.-L. 1988. Mechanical properties of frozen saline clays, *In*: *Proceedings 5th International Conference on Permafrost*, Trondheim, Norway, 1078–1084.

GLASSER, N. & HAMBREY, M. 2001. Styles of sedimentation beneath Svalbard valley glaciers under changing dynamic and thermal regimes. *Journal of the Geological Society London*, **158**, 697–707.

GREGERSEN, O., PHUKAN, A. & JOHANSEN, T. 1983. Engineering properties and foundation design alternatives in marine Svea Clay, Svalbard. *In: 4th International Conferance on Permafrost*, Fairbanks, AK, 384–388.

GRIPP, K. 1929. Glaziologische und geologische Ergebnisse der Hamburgischen Spitzbergen Expedition. *Abhandlungen des Naturwissenschaftlichen Vereins zu Hamburg*, **22**, 147–247.

GUGLIELMIN, M. 2004. Observations on permafrost ground thermal regimes from Antarctica and the Italian Alps, and their relevance to global climate change. *Global and Planetary Change*, **40**, 159–167.

HAEBERLI, W. 1979. Holocene push-moraines in alpine permafrost. *Geografiska Annaler*, **61A**, 43–48.

HAEBERLI, W. 1992. Construction, environmental problems and natural hazards in periglacial mountain belts. *Permafrost and Periglacial Processes*, **3**, 111–124.

HAEBERLI, W., CHENG, G., GORBUNOV, A. P. & HARRIS, S. A. 1993. Mountain permafrost and climatic change. *Permafrost and Periglacial Processes*, **4**(2), 165–174.

HAGEN, J. O. 1978. Brefrontprosesser ved Hardangerjøkulen [Glacier front processes at Hardangerjøkulen]. Cand. real. Thesis, University of Oslo, Norway, Oslo.

HAGEN, J. O. 1988. Glacier surge in Svalbard with examples from Usherbreen. *Norsk Geografisk Tidskrift*, **42**(4), 239–252.

HAGEN, J. O., KOHLER, J., MELVOLD, K. & WINTHER, J. G. 2003. Glaciers in Svalbard: Mass balance, runoff and freshwater flux. Polar Research, **22**(2), 145–160.

HALLET, B., HUNTER, L. & BOGEN, J. 1996. Rates of erosion and sediment evacuation by glaciers: A review of field data and their implications. *Global and Planetary Change*, **12**, 213–235.

HAMBREY, M. J., DOWDESWELL, J. A., MURRAY, T. & PORTER, P. R. 1996. Thrusting and debris entrainment in a surging glacier: Bakaninbreen, Svalbard. *Annals of Glaciology*, **22**, 241–248.

HAMBREY, M. J., BENNETT, M. R., DOWDESWELL, J. A., GLASSER, N. F. & HUDDART, D. 1999. Debris entrainment and transfer in polythermal valley glaciers. *Journal of Glaciology*, **45**(149), 69–86.

HARRIS, C. & BOTHAMLEY, K. 1984. Englacial deltaic sediments as evidence for basal freezing and marginal shearing, Leirbreen, Southern Norway. *Journal of Glaciology*, **30**(104), 30–34.

HARRIS, C., VONDER MÜHLL, D., ISAKSEN, K., HAEBERLI, W., SOLLID, J. L., KING, L., HOLMLUND, P., DRAMIS, F., GUGLIELMIN, M. & PALACIOS, D. 2003. Warming permafrost in European mountains. *Global and Planetary Change*, **39**, 215–225.

HART, J. K. & WATTS, R. J. 1997. A comparison of the styles of deformation associated with two recent push moraines, south Van Keulenfjorden, Svalbard. *Earth Surface Processes and Landforms*, **22**, 1089–1107.

HEGGEM, E. S. F., JULIUSSEN, H. & ETZELMÜLLER, B. 2003. Mountain permafrost in the Sølen massif, Central-Eastern Norway. *In*: PHILLIPS, M.

SPRINGMAN, S. M. & ARENSON, L. U. (eds), *Eighth International Conference on Permafrost, Proc.*, Balkema, Zurich, 367–372.

HODGKINS, R. 1997. Glacier hydrology in Svalbard, Norwegian High Arctic. *Quaternary Science Reviews*, **16**, 957–973.

HODGKINS, R., COOPER, R., WADHAM, J. & TRANTER, M. 2003. Suspended sediment fluxes in a high-Arctic glacierised catchment: Implications for fluvial sediment storage. *Sedimentary Geology*, **162**, 105–117.

HODSON, A. J., TRANTER, M., DOWDESWELL, J. A., GURNELL, A. M. & HAGEN, J. O. 1997. Glacier thermal regime and suspended-sediment yield: A comparison of two high-Arctic glaciers. *Annals of Glaciology*, **24**, 32–37.

HUMLUM, O. 2004. *Holocene Permafrost Aggradation in Svalbard. In*: Harris, C. & Murton, J. (eds.): *Cryospheric Systems: Glaciers and Permafrost*. Geological Society, London, Special Publications, **242**, 119–130.

HUMLUM, O., INSTANES, A. & SOLLID, J. L. 2003. Permafrost in Svalbard: A review of research history, climatic background and engineering challenges. *Polar Research*, **22**(2), 191–215.

HUMLUM, O., ELBERLING, B., HORMES, A., FJORDHEIM, K., HANSEN, O. H. & HEINEMEIER, J. 2005. Late Holocene glacier growth in Svalbard, documented by subglacial find of old vegetation and still alive soil microbes. *The Holocene* (in press).

INTERNATIONAL PERMAFROST ASSOCIATION 1998. *Multi-language Glossary of Permafrost and Related Ground-ice Terms*, IPA, The University of Calgary, Calgary.

ISAKSEN, K., HAUCK, C., GUDEVANG, E., ØDEGÅRD, R. S. & SOLLID, J. L. 2002. Mountain permafrost distribution on Dovrefjell and Jotunheimen, southern Norway, based on BTS and DC resistivity tomography data. *Norsk Geografiska Tidsskrift*, **56**(2), 122–136.

JOHNSON, P. G. 1971. Ice-cored moraine formation and degradation, Donjek glacier, Yukon Territory, Canada. *Geografiska Annaler*, **53A**, 198–202.

KING, L. 1986. Zonation and ecology of high mountain permafrost in Scandinavia. *Geografiska Annaler*, **68A**, 131–139.

KJÆR, K. H. & KRÜGER, J. 2001. The final phase of dead-ice moraine development: Processes and sediment architecture, Kotlujokull, Iceland. *Sedimentology*, **48**(5), 935–952.

KLEMAN, J. 1994. Preservation of landforms under ice sheets and ice caps. *Geomorphology*, **9**, 19–32.

KNEISEL, C. 2003. Permafrost in recently deglaciated glacier forefields – measurements and observations in the eastern Swiss Alps and northern Sweden. *Zeitschrift fuer Geomorphologie*, **47**(3), 289–305.

KRISTIANSEN, K. & SOLLID, J. L. 1987. Svalbard, Glasial geologi og geomorfologi, 1:1000000. Geografisk institutt, Universitetet i Oslo.

Krüger, J. & Kjær, K. H. 2000. De-icing progression of ice-cored moraines in a humid, subpolar climate, Kotlujökull, Iceland. *The Holocene*, **10**(6), 737–747.

LACHENBRUCH, A. H. & MARSHALL, B. V. 1986. Changing climate: Geothermal evidence from permafrost in the Alaskan Arctic. *Science*, **234**, 689–696.

LAWSON, D. E. 1993. Glaciohydrologic and glaciohydraulic effects on runoff and sediment yield in glacierized basins. US Army Corps of Engineers, Cold Region Research and Engineering Laboratory (CRREL), 93-2.

LIESTØL, O. 1965. Lokalt område med permafrost i Gudbrandsdalen. *Norsk Polarinstitutts Aarbok*, **1965**, 129–133.

LIESTØL, O. 1967. Storbreen glacier in Jotunheimen, Norway. *Norsk Polarinstitutt Skrifter*, **141**, 63.

LIESTØL, O. 1977. Pingos, springs, and permafrost in Spitsbergen. *Norsk Polarinstitutt Årbok*, **1975**, 7–29.

LIESTØL, O., 1994. Kompendium i glasiologi, Geografisk institutt, Universitetet i Oslo, Oslo.

LIESTØL, O. 1996. Open-system pingos in Spitsbergen. *Norsk geografisk Tidsskrifts*, **50**, 81–84.

LIESTØL, O. & SOLLID, J. L. 1980. Glacier erosion and sedimentation at Hardangerjøkullen and Omnsbreen. *In*: ORHEIM, O. (ed.), *Symposium on Processes of Glacier Erosion and Sedimentation, Field Guide to Excursion*. Norsk Polarinstitutt, Oslo, Geilo, Norway, 1–22.

LYSÅ, A. & LØNNE, I. 2001. Moraine development at a small High-Arctic valley glacier: Rieperbreen, Svalbard. *Journal of Quarternary Science*, **16**(6), 519–529.

MANGERUD, J., JANSEN, E. & LANDVIK, J. Y. 1996. Late Cenocoic history of the Scandinavian and Barents Sea ice sheets. *Global and Planetery Change*, **12**, 11–26.

MATTHEWS, J. A., McCARROLL, D. & SHAKESBY, R. A. 1995. Contemporary terminal-moraine ridge formation at a temperate glacier – Styggedalsbreen, Jotunheimen, Southern Norway. *Boreas*, **24**(2), 129–139.

NESJE, A., DAHL, S. O., ANDERSSON, C. & MATTHEWS, J. A. 2000. The lacustrine sedimentary sequence in Sygneskardvatnet, western Norway: A continuous, high-resolution record of the Jostedalsbreen ice cap during the Holocene. *Quaternary Science Reviews*, **19**, 1047–1065.

NESJE, A., MATTHEWS, J. A., DAHL, S. O., BERRISFORD, M. S. & ANDERSSON, C. 2001. Holocene glacier fluctuations and winter precipitation changes in the Jostedalsbreen region, western Norway: Evidence from pro-glacial lacustrine sediment records. *The Holocene*, **11**, 267–280.

ØDEGÅRD, R., HAMRAN, S. E., BØ, P. H., ETZELMÜLLER, B., VATNE, G. & SOLLID, J. L. 1992. Thermal regime of a valley glacier, Erikbreen, northern Spitsbergen. *Polar Research*, **11**, 69–80.

ØDEGÅRD, R. S., HOELZLE, M., JOHANSEN, K. V. & SOLLID, J. L. 1996. Permafrost mapping and prospecting in southern Norway. *Norsk geografisk Tidsskrift*, **50**, 41–54.

ØDEGÅRD, R. S., HAGEN, J. O. & HAMRAM, S.-E. 1997. Comparison of radio-echo sounding (30–1000 MHz) and high-resolution borehole-temperature measurements at Finsterwalderbreen, southern Spitsbergen, Svalbard. *Annals of Glaciology*, **24**, 262–271.

ØSTREM, G. 1964. Ice-cored moraines in Scandinavia. *Geografisk Annaler*, **46A**, 282–337.

ØSTREM, G. 1975. Sediment transport in glacial meltwater streams. *In*: JOPLING, A. V. & McDONALD, B. C. (eds) *Glaciofluvial and Glaciolacustrine Sedimentation*, Society of Economic Paleontologists, Tulsa, OK, 101–122.

PATERSON, W. S. B. 1994. *The Physics of Glaciers*, Pergamon/Elsevier Science, Oxford.

SHUMSKIY, P. A. 1964. Principles of structural glaciology: The petrography of fresh-water ice as a method of glaciological investigation. Translated from the Russian by KRAUS, D, Dover, New York.

SLETTEN, K., LYSÅ, A. & LØNNE, I. 2001. Formation and disintegration of a high-arctic ice-cored moraine complex, Scott Turnerbreen, Svalbard. *Boreas*, **30**(4), 272–284.

SNYDER, J. A., WERNER, A. & MILLER, G. H. 2000. Holocene cirque glacier activity in western Spitsbergen, Svalbard: Sediment records from proglacial Linnevatnet. *The Holocene*, **10**(5), 555–563.

SOLLID, J. L. & SØRBEL, L. 1984. Distribution and genesis of moraines in Central Norway. *Striae*, **20**, 63–67.

SOLLID, J. L. & SØRBEL, L. 1988a. Influence of temperature conditions in formation of end moraines in Fennoscandia and Svalbard. *Boreas*, **17**, 553–558.

SOLLID, J. L. & SØRBEL, L. 1988b. Utbredelsesmønsteret av løsmateriale og landformer på Svalbard – noen hovedtrekk. *Norsk geografisk Tidsskrift*, **42**(4), 265–270.

SOLLID, J. L. & SØRBEL, L. 1994. Distribution of glacial landforms in southern Norway in relation to the thermal regime of the last continental ice-sheet. *Geografiska Annaler*, **76**(1–2), 25–35.

SOLLID, J. L. & SØRBEL, L. 1998. Palsa bogs as a climate indicator – examples from Dovrefjell, Southern Norway. *Ambio*, **27**(4), 287–291.

SOLLID, J. L., ETZELMÜLLER, B., VATNE, G. & ØDEGÅRD, R. 1994. Glacial dynamics, material transfer and sedimentation of Erikbreen and Hannabreen, Liefdefjorden, northern Spitsbergen. *Zeitschrift für Geomorphologie*, Suppl. Bind, **97**, 123–144.

SOLLID, J. L., ISAKSEN, K., EIKEN, T. & ØDEGÅRD, R. S. 2003. The transition zone of mountain permafrost on Dovrefjell, southern Norway. *In*: PHILLIPS, M., SPRINGMAN, S. M. & ATENSON, L. U. (eds.), *Eighth International Conference on Permafrost, Proc.*, Balkema, Zurich, 1085–1090.

SOUCHEZ, R. A. 1971. Ice-cored moraines in South-Westen Ellesmere ISLAND, N.W.T, Canada. *Journal of Glaciology*, **10**, 245–254.

SUTER, S., LATERNSER, M., HAEBERLI, W., FRAUENFELDER, R. & HOELZLE, M. 2001. Cold firn and ice of high-altitude glaciers in the Alps:

Measurements and distribution modelling. *Journal of Glaciology*, **47**(156), 85–96.

SVENDSEN, J. I. & MANGERUD, J. 1997. Holocene glacial and climate variations on Spitsbergen, Svalbard. *The Holocene*, **7**(1), 45–57.

SVENDSEN, J. I., MANGERUD, J. & MILLER, G. H. 1989. Denudation rates in the Arctic estimated from lake sediments on Spitsbergen, Svalbard. *Palaeogeography, Palaeoclimatology, Palaeoecology*, **76**(1–2), 153–168.

TSYTOVICH, N. A. 1975. *The Mechanics of Frozen Ground*, McGraw-Hill, Washington, DC.

VATNE, G., ETZELMÜLLER, B., ØDEGÅRD, R. & SOLLID, J. L. 1995. Subglacial drainage and sediment evacuation at Erikbreen, northern Spitsbergen. *Nordic Hydrology*, **26**, 169–190.

WEERTMAN, J. 1961. Mechanism for the formation of inner moraines found near the edge of cold ice caps and ice sheets. *Journal of Glaciology*, **3**, 965–978.

WILLIAMS, P. J. & SMITH, M. W. 1989. *The Frozen Earth: Fundamentals of Geocryology*, Cambridge University Press, Cambridge.

Investigating glacier–permafrost relationships in high-mountain areas: historical background, selected examples and research needs

WILFRIED HAEBERLI

*Glaciology and Geomorphodynamics Group, Geography Department,
University of Zurich, Zurich, Switzerland*

Abstract: Investigations on the relationships and interactions between glaciers and permafrost in high-mountain regions have long been neglected. As a consequence, numerous fascinating questions remain open and offer possibilities for highly relevant, innovative and integrative research concerning materials, processes, landforms, environmental aspects and natural hazards. The historical background to this situation is first reviewed, examples are given of some key unanswered questions and two case studies are presented to illustrate the importance of considering the combined effects of glaciers and permafrost, particularly in the context of hazard assessments in high mountains.

Introduction and historical background

For historical reasons, research on glaciers and permafrost has primarily evolved along separate lines. Permafrost science has its main roots in glacier-free sub-Arctic and Arctic lowlands, whereas the science of glaciers originated in high mountains where permafrost is not easily recognized. Today, diverging scientific cultures exist in the two fields with physicists largely influencing glacier research and geoscientists/ engineers leading permafrost science. Corresponding terminological discrimination and confusion in various countries mirrors this unfortunate situation (e.g., 'geocryology' or 'cryopedology' vs 'glaciology', with the latter term being incorrectly used for glacier research only). In fact, this dichotomy within the snow-and-ice community limits its credibility with respect to other domains and requires special coordinating efforts in order to avoid linear thinking and omission of fundamental aspects. Considerable shortcomings have developed, especially in the progress of snow and ice research in high-mountain areas, where relationships and interactions between snow, glaciers and permafrost have hardly been investigated outside Europe. The lack of mutual knowledge and interest between the subdisciplines may lead to serious misconceptions and, especially in relation to environment problems or natural hazards, outcomes may have devastating effects. Such shortcomings, on the other hand, also represent major opportunities for innovative

research connecting knowledge and understanding from all sides.

Modern efforts concerning serious challenges, such as global climate change, global environment-related observation, radioactive waste disposal in formerly glaciated or frozen terrain and hazard mitigation in cold regions, increasingly call for integrated views of the cryosphere components and related processes within the earth system. However, integrated glacier/ permafrost concepts have been developed by only a few scientists. Outstanding examples are Shumskii (1964), with his textbook *Principles of Structural Glaciology* on ice formation in glacial and periglacial areas, or Liestøl (1989) with his extensive experience of Svalbard concerning thermal and hydraulic conditions in coal mining areas beneath polythermal glaciers with sub-glacial permafrost. These authors provided the basis for our understanding that regional occurrences of temperate/polythermal/ cold glaciers and sub-/peri- glacial permafrost are largely a function of topography and continentality of the climate (Fig. 1). Recent research on the interactions between glacial and permafrost domains is providing new insights, and some examples may illustrate the benefits of, and challenges for, this long-neglected aspect of high mountain glaciology and geomorphology. Selected examples are presented below concerning landforms, materials, processes, and environments in relation to natural hazards in mixed glacier/permafrost areas.

From: HARRIS, C. & MURTON, J. B. (eds) 2005. *Cryospheric Systems: Glaciers and Permafrost.*
Geological Society, London, Special Publications, **242**, 29–37.
0305-8719/05/$15.00 © The Geological Society of London 2005.

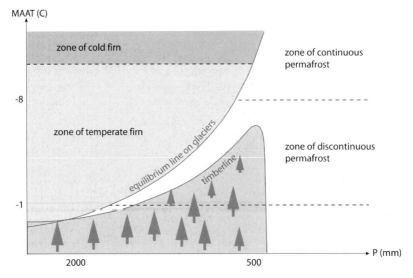

Fig. 1. Cryosphere model that illustrates the spatial relations between glaciers and permafrost as a function of mean annual air temperature and annual precipitation (from Haeberli & Burn 2002)

Examples of recent research on glacier–permafrost interaction

Materials and processes: ice formation with glacier–permafrost contacts

Formation, preservation or melting of permafrost underneath and in front of polythermal mountain glaciers is a complex topic that is still far from being understood. Recent studies using modern field methodologies are presented by Kneisel *et al.* (2000), Kneisel (2003) or Etzelmüller & Hagen (2005) for European mountains. Latent heat in sub-glacial permafrost greatly influences basal thermal regimes and consequently the flow of polythermal glaciers and ice caps. Such effects are, however, seldom taken into account in numerical flow models of such surface ice bodies. The formation of basal regelation layers is nothing other than frost penetration into non-frozen sediments with corresponding production of interstitial and segregation ice. This leads to ice saturation and supersaturation of the frozen and subsequently entrained layer. The rheology of this new 'dirty' ice base corresponds to the mechanical characteristics of frozen ground (Fig. 2) and raises serious questions concerning the isotropy assumption made in most, if not all, glacier and ice-sheet models. Depending on the degree of saturation, the deformation will follow a damped or steady-state creep mode, causing a 'stiff' or 'flexible' contact with the underlying ground. The resulting difference between basal friction of a ductile layer consisting of 'ice with

debris' vs basal friction of a rigid layer consisting of 'debris with ice' influences glacier–bed coupling, overall flow stability ('continuous' vs 'stop-and-go' sliding) as well as abrasion rates (Schweizer & Iken 1992).

Landforms: rock glaciers and glacitectonically deformed moraines

Speculation about the origins of rock glaciers ('are they glaciers or permafrost?') may be the most striking case of interpreting cryogenic phenomena as being exclusively glacial or permafrost in origin, as if there were no possibility for the two to coexist or interact. The idea that lava stream-like rock-glacier bodies can develop over millennia from glaciers alone (i.e., in the absence of permafrost) may be primarily the outcome of: (a) the non-consideration of negative ground temperatures and the ice-formation phenomena involved in terms of spatio-temporal scales and processes; (b) the tacit assumption that all surface ice is temperate (at pressure-melting temperature); and (c) the confusion between flowing glaciers and more or less static perennial snow-banks or glacierets with their specific thermo-mechanical properties (Haeberli 2000). Correct use of basic knowledge from both glacier and permafrost science, together with the use of adequate exploration and modelling techniques, immediately opens the way to sound concepts and advanced understanding of these fascinating

Fig. 2. Regelation layer (scale about 1 m) at the base of the (then) advancing Rhone glacier – a supersaturated perennially frozen layer deforms in a ductile mode at the base of relatively pure 'glacier' ice above. Photograph by the author, around 1990

phenomena (cf. many papers in ICOP 2003). Based on such modern approaches, the creep of perennially frozen, ice-rich debris under the influence of gravity alone (e.g., 'rock glacier') or via the influence of gravity-driven flowing cold or polythermal mountain glaciers (e.g., push morianes *sensu stricto*, cf. Haeberli 1979; Kälin 1971) is now quite well documented by drilling, geophysics and photogrammetry. Only appropriate consideration of both polythermal glaciers and permafrost conditions makes it possible to fully understand the polygenetic nature of ground ice in such situations. Likewise, accurate assessment of the complex suite of landforms from deformed frozen talus to periglacial rock glaciers, deformed ice-containing moraines and polythermal to cold (sometimes debris-covered) glaciers (Fig. 3) requires careful consideration of the geothermal regime and glacial history of the site. This is especially important in the context of climate change, environmental aspects and hazard assessments in glacier–permafrost regions.

Environmental aspects and hazards: radioactive waste disposal, geothermal energy, lake outbursts and ice/rock avalanches

Not only the Pleistocene Scandinavian ice sheet formerly covering a vast mountainous terrain but also the large Pleistocene piedmont glaciers that descended from the highest summits of the Alps into the perennially frozen lowlands of central Europe (Frenzel *et al.* 1992) have been polythermal with cold parts dominating during the coldest/dryest periods (Blatter & Haeberli 1984; Boulton & Payne 1994). Such conditions probably caused deep effects on the thermal and hydraulic field of the uppermost crust (hundreds of metres to a few kilometres; Boulton & Caban 1995; Haeberli 1981; Haeberli *et al.* 1984; Haeberli & Schlüchter 1987; Speck 1994) and relate directly to the present-day regional patterns of ground heat available for geothermal energy production and to the future long-term safety of deeply buried radioactive waste. Understanding these effects requires sophisticated integration of glacier and permafrost models (e.g., Boulton & Caban 1995; Boulton & Hartikainen 2003).

Glacier–permafrost interactions are also essential aspects to be included in research and assessments of high-mountain hazards. This especially concerns the stability of ice-marginal and thermokarst lakes in periglacial mountain belts, of steep cold or polythermal hanging glaciers, and of slopes with degrading permafrost and shrinking glaciers inducing deep-seated changes in stress conditions and in mechanical/hydraulic properties (Evans & Clague 1988). Cold glacier margins frozen to their bed may not float if marginal lake levels rise, enabling lake drainage by overflow – an often much less

Fig. 3. Retreating glacier tongue in the Tien Shan Mountains of Kazakhstan showing ice-cored moraines, a proglacial lake and traces of an earlier and now empty lake. The 'swollen' and viscous appearance of the moraines indicates creep within perennially frozen ice-rich material. The potential of such pro- and peri-glacial lakes to produce sudden outbursts, breach the moraines and trigger large debris flows is directly connected to the presence/absence of permafrost. Photograph by Stephan Gruber, 2001

dangerous process than progressive enlargement of subglacial channels (Maag 1967; cf. Fig. 4). Melting of massive ice (thermokarst) within non-consolidated subglacial permafrost exposed by retreating polythermal glaciers with cold margins can lead to extreme porosities and even cavities (Vonder Mühll *et al.* 1996). This process favours piping from proglacial moraine lakes and subsequent formation of debris flows (cf. Haeberli *et al.* 2001; Huggel *et al.* 2003). Hanging glaciers with relatively warm firn surfaces can induce deep thermal anomalies in perennially frozen rock walls (Haeberli *et al.* 1997) and can be related to major ice/rockfall events such as the recent catastrophe in the Caucasus (Haeberli *et al.* 2003). The extreme heat during summer 2003 in central Europe may have induced corresponding long-term thermal effects on the highest glacier-clad summits of the Alps. Such thermal disturbances combined with unloading of slopes due to vanishing glaciers may greatly influence the scaling of paraglacial mass exhaustion processes (Ballantyne 2003). The complexity of permafrost–glacier interactions and the fundamental need to consider both in combination when assessing hazards in high mountains is illustrated by the case of hazardous lakes in the cirque at Gruben in the Swiss Alps and of the starting zone for the large Kolka-Karmadon rock/ice slide (Osetian Caucasus, Russia).

Case Histories of glacier–permafrost Interaction

Combined glacier–permafrost effects at the Gruben Lakes, Switzerland

In the Gruben cirque, a polythermal glacier and the relatively warm permafrost of a large, active rock glacier are juxtaposed. Long-term research in this area has accompanied and assisted successful flood prevention work from ice-marginal as well as thermokarst lakes after two major lake outbursts and debris flows in 1968 and 1970 (Haeberli *et al.* 2001). Figure 5 presents the site and its most critical aspects. The highest reaches of the accumulation area of Gruben glacier (1 on Fig. 5) are cold and frozen to the bedrock beneath (Haeberli 1976; Suter *et al.* 2001). Such cold firn areas are likely to be warming with rising air temperatures, inducing faster flow (enhanced deformation and sliding) of the uppermost part of the glacier. A recent rock fall (2 on Fig. 5) started from perennially frozen rock at about 3600 m (a.s.l.; above sea level; Noetzli *et al.* 2003; cf. Davies *et al.* 2001) and travelled along the centreline of the glacier surface to an altitude of about 3100 m a.s.l. Such mass movements across glaciers can have especially long flow paths and a major rockfall could possibly reach the area of the ice-marginal lakes and the flood protection structures. This is

(a)

(b)

Fig. 4. Lake 3 at Gruben before and after the construction of an overflow channel (circle in second image). Floating of the ice dam was probably prevented by the fact that the glacier margin is frozen to sub-glacial permafrost and water cannot penetrate directly to the glacier bed. Artificial lowering of the ice-marginal lake was necessary, because the retention capacity designed earlier for proglacial Lake 1 was not sufficient for combinations of events such as heavy precipitation or a subglacial water-pocket rupture combined with an outburst of Lake 3. Photographs by Andreas Kääb, 29 July and 7 August 2003

especially true for the permafrost rock wall of the Inner Rothorn (3 on Fig. 5), where intense rock fall activity provides larger blocks (metre to deca-metre scale), as must be expected from seasonal frost weathering and permafrost degradation (Matsuoka *et al.* 1998). Lowering of the glacier surface at this site introduces stress redistribution within the rock face combined with greater frost penetration and intensified rock destruction (cf. Haeberli *et al.* 1997). Thus, this is a location with rapidly changing stability conditions. The now-regulated so-called 'Lake 3' (4 on Fig. 5) continues to enlarge towards the ice dam of the thinning glacier, and in 2003 it reached the floata-tion level of the ice dam, which is cold and still frozen to the subglacial permafrost at the margin of the polythermal glacier. An artificial cut in the ice dam helped to accelerate and enhance overflow. The lake did not drain completely and must be kept under continuous

Fig. 5. Overview of the situation at Gruben where permafrost–glacier interactions determine hazards induced by climate change (numbers refer to the explanation in the text; the moraine dam at Lake 1 with the breach of earlier outbursts is partially hidden below the wing of the aeroplane). Photograph (taken from the air) by Christine Rothenbühler October 2003

observation, especially if it grows towards the vanishing glacier tongue and the rock-fall zone. The permafrost of the rock glacier (5 on Fig. 5) was in contact with the glacier during earlier (Holocene and historical) advances. It still contains buried massive ice, which is a favourable precondition for the exponential growth of potentially dangerous thermokarst lakes (Kääb & Haeberli 2001). The moraine dam at 'Lake 1' (6 on Fig. 5) is now protected by a boulder dam, a specially designed outlet structure and concrete injections. At this location, breaching in non-consolidated material with large cavities from former subglacial permafrost took place during two outburst floods in 1968 and 1970.

Combined glacier–permafrost effects in the starting zone of the Kolka-Karmadon rock/ice avalanche, Osetian Caucasus, Russia

The large rock/ice slide of 20 September 2002 from the north-northeast wall of the summit of Dzhimarai-khokh, Kazbek massif, Northern Osetia, Russian Caucasus, killed more than 140 people and destroyed the access road through the Giseldon Valley and Genaldon gorge – a

primary tourist attraction of the region (Kääb *et al.* 2003; Popovnin *et al.* 2003). The avalanche starting zone as depicted in Fig. 6 was approximately 1 km wide and located between about 4300 and 3500 m a.s.l. A rough estimation of the volume of steeply inclined metamorphic rock layers detached in the slide is some 4 million m^3, with failure extending to a depth of about 40 m. A similar thickness/volume of snow, firn and glacier ice was also entrained in the slide. The primary cause of the instability must therefore have been within bedrock rather than surface ice. In view of the fact that bedrock stability in cold mountain areas can be especially low in warm or degrading permafrost (Davies *et al.* 2001), thermal conditions affecting ice and water within rock joints are likely to have exerted a major and detrimental influence.

The situation after the event necessitated an immediate assessment of the potential for a repetition of similar or even larger accidents from the mountain slope. This involved interpretation of photographs collected by the authorities during reconnaissance flights by helicopter together with some best guesses about the thermal condition of firn, ice and permafrost in the starting zone (Haeberli *et al.* 2003). Bedrock surface temperatures in the detachment zone

Fig. 6. Upper avalanche path of the Kolka-Karmadon rock/ice slide. The summit of Dzhimarai-khokh and the starting zone (S) are in the background. An active talus-derived rock glacier (R) is seen in the lower right corner. Note the steam/dust (?) cloud at the foot of the slope where the Kolka glacier (K) has been sheared off. Two flow parts (F1, F2) of the fast travelling rock/ice/water mass can be discriminated. The Kazbek volcano would be to the left. Photograph by Igor Galushkin 25 September 2002

were estimated to be about -5 to $-10°C$, i.e., conditions of cold permafrost. Before the event, the steep impermeable lower rock slope had been covered by hanging glaciers. Such hanging glaciers have two thermally different parts: (1) cold ice frozen to bedrock forms the vertical/impermeable ice cliffs where meltwater runs off immediately and ice lamellas break off as an important ablation process of these ice bodies, while percolation and refreezing of meltwater cause the existence of (2) much warmer or even temperate firn and ice below the less steep upper surfaces where snow accumulation predominates. A polythermal structure is likely to have existed in the hanging glaciers, at least in the lower parts of the wall (cf. Haeberli *et al.* 1997). The detachment zone at Dzhimarai-khokh had therefore probably been in a complex condition of relatively cold/thick permafrost combined with warm if not unfrozen parts with meltwater flow in very steeply inclined materials with heterogenous permeability. This situation favoured high and locally variable water pressures and was further complicated by the fact that hot springs are known to occur in this volcanic region of the Kazbek massif. The event itself removed hanging glaciers with warm firn areas, reducing the load on the slope and eliminating the main meltwater source. Moreover, the exposed bedrock is now subject

to strong cooling and deep freezing. These two facts led to the conclusion that the threat of a similar or even bigger event at the same site in the immediate future could be considered minimal but that instabilities in the remaining parts of the slope could continue and should be observed accordingly. Effects of potential future atmospheric warming would also have to be taken into account over the coming decades.

Suggested research needs and recommendations

Glaciers and permafrost in cold mountain areas often coexist and give rise to important complex phenomena and processes that can only be understood by considering the thermal regime of both the glacier ice and the underlying substrate. Fundamental concepts of ice formation above and below the earth surface must be understood, together with the spatio-temporal scales in the variation of sub-zero ice and ground temperatures. Assessments of climate change effects, or of hazards to human installations in mountain areas, require consideration of both frozen ground and perennial surface ice, together with their complex interactions and changes through time.

The two sub-disciplines, glacier science and permafrost science, have potential for major integrative research. This 'new scientific land' between the two domains should encourage studies of glacier–permafrost relationships and interactions. The primary challenge is to overcome the historical barriers that exist between the two disciplines and to integrate rather than exclude knowledge and understanding. Closer cooperation would also be advisable with respect to institutional structures. The International Commission on Snow and Ice (ICSI) may attempt to become an association within the International Union of Geodesy and Geophysics (IUGG), but must include the science of frozen ground in order to do justice to the entire scientific field that it supports. The International Permafrost Association, on the other hand, should broaden its scope in order to make the scientific contributions needed to advance our understanding of the glacier–permafrost interface. The establishment of a combined 'International Cryosphere Association' or 'International Association for Snow, Ice and Permafrost' is being recommended from various sides. The initiation of an ICSI/IPA working group on glacier and permafrost hazards in mountain areas is an important step to this direction.

The author extends his thanks to all those colleagues who have strengthened his interest in glacier–permafrost relationships and helped develop corresponding concepts and approaches through sharing of experience, stimulating discussions and encouraging attitude. Special thanks are due to Charles Harris, Julian Murton and an anonymous reviewer for their helpful comments on an earlier version of the present contribution.

References

BALLANTYNE, C. K. 2003. Paraglacial landform succession and sediment storage in deglaciated mountain valleys: theory and approaches to calibration. *Zeitschrift für Geomorphologie*, **132** (Suppl.), 1–18.

BLATTER, H. & HAEBERLI, W. 1984. Modelling temperature distribution in Alpine glaciers. *Annals of Glaciology*, **5**, 18–22.

BOULTON, G. S. & CABAN, P. 1995. Groundwater flow beneath ice sheets: part II – its impacts on glacier tectonic structures and moraine formation. *Quaternary Science Reviews*, **14**, 563–587.

BOULTON, G. & HARTIKAINEN, J. 2003. Thermo-hydro-mechanical impacts of coupling between glaciers and permafrost. *In: GeoProc*, Stockholm, 283–288.

BOULTON, G. S. & PAYNE, T. 1994. Mid-latitude ice sheets through the last glacial cycle: glaciological and geological reconstructions. *NATO ASI Series*, **122**, 177–212.

DAVIES, M., HAMZA, O. & HARRIS, C. 2001. The effect of rise in mean annual air temperature on the stability of rock slopes containing ice-filled discontinuities. *Permafrost and Periglacial Processes*, **12**, 137–144.

ETZELMÜLLER, B. & HAGEN, J. O. 2005. Glacier–permafrost interaction in Arctic and alpine mountain environments with examples from southern Norway and Swalbard. *In*: HARRIS, C. & MURTON, J. B. (eds.): *Cryospheric Systems: Glaciers and Permafrost*. Geological Society, London, Special Publications, **242**, 11–27.

EVANS, S. G. & CLAGUE, J. J. 1988. Catastrophic rock avalanches in glacial enviroments. Landslides. *In: Proc. of the fifth International Symposium on Landslides*, Lausanne, 10–15 July 1988, Vol. 2, 1153–1158.

FRENZEL, B., PÉCSI, M. & VELICHKO, A. A. 1992. *Atlas of Paleoclimates and Paleoenvironments of the Northern Hemisphere – Late Pleistocene/Holocene*, Hungarian Academy of Sciences and Gustav Fischer Verlag, Budapest and Stuttgart.

HAEBERLI, W. 1976. Eistemperaturen in den Alpen. *Zeitschrift für Gletscherkunde und Glazialgeologie*, **XI/2**, 203–220.

HAEBERLI, W. 1979. Holocene push-moraines in Alpine permafrost. *Geografiska Annaler*, **61A**(1–2): 43–48.

HAEBERLI, W. 1981. Ice motion on deformable sediments. *Journal of Glaciology*, **27**(96), 365–366.

HAEBERLI, W. 2000. Modern research perspectives relating to permafrost creep and rock glaciers. *Permafrost and Periglacial Processes*, **11**, 290–293.

HAEBERLI, W. & BURN, C. 2002. Natural hazards in forests – glacier and permafrost effects as related to climate changes. *In*: SIDLE, R. C. (ed.) *Environmental Change and Geomorphic Hazards in Forests*, IUFRO Research Series, Vol. 9, 167–202.

HAEBERLI, W. & SCHLÜCHTER, C. 1987. Geological evidence to constrain modelling of the Late Pleistocene Rhonegletscher, Swiss Alps. The physical basis of ice sheet modelling. *In: Proc. Vancouver Symposium*, August 1987, IAHS Publication no. 170, 333–346.

HAEBERLI, W., RELLSTAB, W. & HARRISON, W. D. 1984. Geothermal effects of 18 ka BP ice conditions in the Swiss Plateau. *Annals of Gaciology*, **5**, 56–60.

HAEBERLI, W., WEGMANN, M. & VONDER MUEHLL, D. 1997. Slope stability problems related to glacier shrinkage and permafrost degradation in the Alps. *Eclogae geologicae Helvetiae*, **90**, 407–414.

HAEBERLI, W., KÄÄB, A., VONDER MÜHLL, D. & TEYSSEIRE, PH. 2001. Prevention of outburst floods from periglacial lakes at Grubengletscher, Valais, Swiss Alps. *Journal of Glaciology*, **47**(156): 111–122.

HAEBERLI, W., HUGGEL, C., KÄÄB, A., POLKVOJ, A., ZOTIKOV, I. & OSOKIN, N. 2003. Permafrost conditions in the starting zone of the Kolka-Kamadon rock/ice slide of 20 September 2002 in

North Osetia (Russian Caucasus). *In*: HAEBERLI, W. & BRANDOVÁ, D (eds). *ICOP 2003 Permafrost: Extended Abstracts Reporting Current Research and New Information. Eighth International Conference on Permafrost*, Zurich, 21–25 July 2003, 49–50.

HUGGEL, C., KÄÄB, A., & HAEBERLI, W. 2003. Regional-scale models of debris flows triggered by lake outbursts: the June 25, 2001 debris flow at Täsch (Switzerland) as a test study. *In*: RICKENMANN, D. & CHEN, C. (eds) *Debris-Flow Hazards Mitigation: Mechanics, Prediction and Assessment*, Millpress, Rotterdam.

ICOP 2003. *Proc. Eighth International Conference on Permafrost*, Zurich, 21–25 July 2003, PHILLIPS, M., SPRINGMAN, S. M. & ARENSON, L. U. (eds). Balkema, Rotterdam, Vols 1 and 2.

KÄÄB, A. and HAEBERLI, W. 2001. Evolution of a high-mountain thermokarst lake in the Swiss Alps. *Arctic, Antarctic and Alpine Research*, **33/4**, 385–390.

KÄÄB, A., WESSELS, R., HAEBERLI, W., HUGGEL, C., KARGEL, J. S. & KHALSA, S. J. S. 2003. Rapid Aster imaging facilitates timely assessments of glacier hazards and disasters. *EOS*, **13**(84), 117, 121.

KÄLIN, M. 1971. The active push moraine of the Thompson Glacier, Axel Heiberg Island, Canadian Arctic Archipelago, Canada. Dissertation 4671, McGill University, Montreal.

KNEISEL, CHR. 2003. Permafrost in recently deglaciated glacier forefields – measurements and observations in the eastern Swiss Alps and northern Sweden. *Zeitschrift für Geomorphologie N. F.*, **47**(3), 289–305.

KNEISEL, CHR., HAEBERLI, W. & BAUMHAUER, R. 2000. Comparison of spatial modelling and field evidence of glacier/permafrost-relations in an Alpine permafrost environment. *Annals of Glaciology*, **31**, 269–274.

LIESTØL, O. 1989. Kompendium i Glasiologi. Meddelelser fra Geografisk institutt Universitetet i Oslo, Naturgeografisk serie. Report no. 15.

MAAG, H. U. 1967. Ice-dammed lakes and marginal glacial drainage on Axel Heiberg Island, Canadian Arctic Artchipelago, Canada. Axel Heiberg Island Research Report, McGill University, Montreal.

MATSUOKA, N., HIRAKAWA, K., WATANABE, T., HAEBERLI, W. & KELLER, F. 1998. The role of diurnal, annual and millennial freeze-thaw cycles in controlling alpine slope instability. *In*: *Proc. Seventh International Conference on Permafrost*, Yellowknife, Canada, Collection Nordicana, Vol. 57, 711–717.

NOETZLI, J., HOELZLE, M. & HAEBERLI, W. 2003. Mountain permafrost and recent Alpine rock-fall events: a GIS-based approach to determine critical factors. *In*: *ICOP 2003 Permafrost: Proc. Eighth International Conference on Permafrost*, Zurich, 21–25 July 2003, Balkema, Rotterdam, Vol. 2, 827–832.

POPOVNIN, V. V., PETRAKOV, D. A., TOUTOUBALINA, O. V. & TCHERNOMORETS, S. S., 2003. Catastrophe glaciaire 2002 en Ossetie du Nord. Unpublished French translation, original article to appear in *Earth Cryosphere*, VII (1).

SCHWEIZER, J. & IKEN, A. 1992. The role of bed separation and friction in sliding over an undeformable bed. *Journal of Glaiology*, **38** (128), 77–92.

SHUMSKII, P. A. 1964. *Principles of Structural Glaciology*, KRAUS, D. (transl), Dover, New York.

SPECK, C. 1994. Änderung des Grundwasserregimes unter dem Einfluss von Gletschern und Permafrost. Dissertation ETHZ, Zürich.

SUTER, S., LATERNSER, M., HAEBERLI, W., FRAUENFELDER, R. & HOELZLE, M. 2001. Cold firn and ice of high-altitude glaciers in the Alps: measurements and distribution modelling. *Journal of Glaciology*, **45**(156), 85–96.

VONDER MÜHLL, D., HAEBERLI, W. & KLINGÉLÉ, E. 1996. Geophysikalische Untersuchungen zur Struktur und Stabilität eines Moränendammes am Grubengletscher (Wallis). *Tagungspublikation Interpraevent, Garmisch-Partenkirchen*, **4**, 123–132.

Glacier–permafrost interactions and glaciotectonic landform generation at the margin of the Leverett Glacier, West Greenland

RICHARD I. WALLER[1] & GEORGE W. TUCKWELL[2]

[1]*School of Earth Sciences and Geography, Keele University, Keele, Staffordshire, ST5 5BG, UK*

(e-mail: r.i.waller@esci.keele.ac.uk)

[2]*STATS Ltd, Poterswood House, Porters Wood, St Albans AL3 6PQ, UK*

Abstract: This paper describes the key characteristics of a proglacial moraine complex at the Leverett Glacier, western Greenland. The presence of a large stream-cut exposure allowed the examination of its internal structure, as well as its surface geomorphology. It is composed of a variety of ice and sediment facies, including debris-poor ice, ice-rich diamicton and ice-rich gravel. These units are glaciotectonized, with the exposure featuring a major fault and associated drag fold, a planar, erosional unconformity, and a variety of small-scale folds. Various interpretations are considered, including the possibility that the sequence represents a buried basal ice layer. However, it is argued that the structural characteristics are best explained by a two-phase model involving ice advance and proglacial or ice-marginal compression, followed by overriding and subglacial deformation and erosion, tentatively related to ice advance after the Holocene Hypsithermal (*c.* 4900–3000 calendar years BP). The polygenetic origin of this ice-marginal, glaciotectonic landform contrasts with the majority of Arctic push-moraines, which are largely considered the result of proglacial deformation and the stacking of imbricate thrust sheets of frozen sediment. This contrast may reflect differences in the thickness and spatial continuity of permafrost within the glacier foreland, and adds to the range of ice-marginal landforms associated with glacier–permafrost interactions.

Interactions between glaciers and permafrost have until recently been considered unimportant for two main reasons. Firstly, cold-based glaciers (with a basal temperature below the pressure melting point) have traditionally been assumed to be slow moving and geomorphologically ineffective, due to the cessation of both basal sliding and subglacial sediment deformation (e.g., Kleman 1994; Paterson 1994; Dowdeswell & Siegert 1999). Secondly, numerical ice-sheet models have suggested that glacier–permafrost interactions are limited in extent, being restricted largely to the marginal fringes of ice masses (e.g., Mooers 1990).

Recent research in both modern-day and Quaternary glacial environments suggests that the importance and potential glaciological significance of glacier–permafrost interactions has been underestimated. Field investigations at cold-based glaciers in Antarctica, Tibet and Iceland have demonstrated how both basal sliding and subglacial sediment deformation can remain active at sub-freezing temperatures

(Echelmeyer & Zhongxiang 1987; Cuffey *et al.* 1999; Bennett *et al.* 2003). In combination with research indicating active sediment entrainment (Cuffey *et al.* 2000), erosion and deposition (Atkins *et al.* 2002) under such conditions, this has led to a reappraisal of the geomorphological effectiveness of cold-based glaciers (Ftizsimons 1996; Fitzsimons *et al.* 2001; Atkins *et al.* 2002). Numerical modelling of ice-sheet lobes also suggests that subglacial permafrost may have been both extensive and long-lived beneath advancing ice. Cutler *et al.* (2000), for example, calculate that subglacial permafrost extended for 60–200 km up-glacier of the margin of the Green Bay lobe of the Southern Laurentide Ice Sheet, potentially remaining intact for up to a few thousand years. This prediction has been supported by field observations in Western Siberia (Astakhov *et al.* 1996) and the Western Canadian Arctic (Murton *et al.* 2005), where extensive subglacial deformation of pre-existing permafrost is described, indicating widespread glacier–permafrost interactions

From: HARRIS, C. & MURTON, J. B. (eds) 2005. *Cryospheric Systems: Glaciers and Permafrost.*
Geological Society, London, Special Publications, **242**, 39–50.
0305-8719/05/$15.00

beneath the high-latitude regions of major Quaternary ice-sheets.

The influence of ice-marginal permafrost on landform development has long been acknowledged. The model of Moran *et al.* (1980), for example, proposed a zonation of glacial landforms, with: (1) an interior zone of thawed-bed conditions characterized by streamlined bedforms (principally drumlins); and (2) a peripheral fringe of frozen-bed conditions characterized by widespread hummocky terrain and tunnel valleys, associated with the thrusting of subglacial debris and the channelization of subglacial drainage, respectively. This model, associated with the southern margin of the Laurentide Ice Sheet in particular, has subsequently been extended and developed by a number of authors (e.g., Mooers 1990; Clayton *et al.* 2001).

In terms of specific landforms, glacier–permafrost interactions are frequently associated with the genesis of push-moraines. These features are widely reported in both modern-day and Quaternary glacial environments, and variously related to the ice-marginal thrusting of imbricate stacks of rock or frozen sediment (in which case they are often termed 'thrust-moraines' or 'thrust-block moraines', e.g., Mackay 1959; Mooers 1990; Evans & England 1991; Krüger 1996), englacial thrusting (e.g., Huddart & Hambrey 1996) and the propagation of glacier-induced stresses into the proglacial area (e.g., Boulton *et al.* 1999). Whilst push-moraines are not restricted to permafrost areas, the largest, most spectacular examples are frequently found around High Arctic glaciers terminating in permafrost (e.g., Kälin 1971; Lehmann 1992; Etzelmüller *et al.* 1996; Hart & Watts 1997; Boulton *et al.* 1999). The presence of proglacial permafrost is indeed considered by some to be central to the development of large, broad push-moraine complexes in particular as it: (1) alters the rheology of proglacial sediments, allowing the transmission of stresses over wider areas; and (2) promotes the generation of high porewater pressures at the base of the permafrost layer and the formation of a décollement surface (Boulton *et al.* 1999).

The assertion that ice-marginal permafrost is a key control upon the formation of large, multi-crested push moraines, and that such features are therefore indicative of glacier–permafrost interactions, remains the subject of ongoing controversy (Bennett 2001). This paper aims to contribute to this debate by describing the geomorphology, composition and structural geology of a large moraine complex at the margin of the Leverett Glacier in western Greenland, and evaluating the role played by glacier–permafrost interactions in its formation.

The Leverett Glacier

The Leverett Glacier is a small outlet glacier in west Greenland, located *c*. 15 km east of Kangerlussuaq (67° 06′ N, 50° 09′ W; Fig. 1). The glacier comprises a southerly offshoot of the larger Russell Glacier, and in combination they constitute a large lobe discharging ice from the western sector of the Greenland ice sheet. The glacier terminates in a region of continuous permafrost, with permafrost only being absent beneath large lakes. The estimated permafrost thickness is 127 ± 31 m, whilst the maximum active-layer thickness at the glacier margin is estimated to be between 0.1 and 2.5 m (van Tatenhove 1995). The glacier has experienced significant oscillations during the last century, having retreated by 200–350 m between 1943 and 1968 and subsequently re-advanced by 325–375 m between 1968 and 1992 (van Tatenhove 1995). Consequently, the glacier represents an ideal site at which to investigate glacier–permafrost interactions.

The current ice margin is characterized by an extensive basal ice layer similar in character to that of the nearby Russell Glacier (see Knight *et al.* 1994; Waller *et al.* 2000 for details) and is bordered by a terminal moraine composed of sandy diamicton *c*. 5–10 m in height (Fig. 2). The proximal proglacial area is dominated by an arcuate moraine complex measuring *c*. 900 m (N–S) by 400 m (E–W; Figs 2 and 3). The complex rises 15–20 m above the surrounding area and is bordered on its proximal and distal sides by small sandur plains. The moraine complex is overlain by a series of smaller ridges 1–2 m high, that mirror its general orientation, and is also punctuated by a series of steep-sided ponds up to *c*. 100 m in diameter. The proximal side of the moraine complex features a conspicuous conical mound *c*. 10 m high, which has been interpreted as a small,

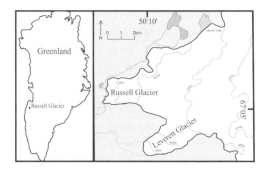

Fig. 1. Location map of the Russell and Leverett Glaciers, west Greenland

Fig. 2. Overview of the proglacial area, indicating key geomorphological features of the Leverett Glacier

open-system pingo (Scholz & Baumann 1997). van Tatenhove (1995) suggested that the moraine complex was created by a major glacier advance following the Holocene Hypsithermal (*c*. 4900–3000 calendar years BP) that terminated at the end of the nineteenth century.

The moraine complex is bordered on its northern side by a major meltwater river that discharges from a subglacial tunnel on the glacier's northern flank. Field investigations at the site in September 2000 revealed a large

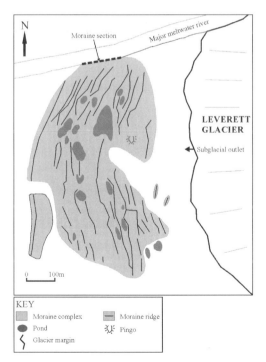

Fig. 3. Geomorphological map of the proglacial area

section eroded through the northern extremity of the moraine complex (Figs 2 and 3). This section forms the focus of this paper.

Moraine section–description

Figure 4 shows a photo-montage of the entire moraine section. The section is orientated roughly transverse to the moraine complex and parallel to the likely direction of ice flow. In 2000, the section measured *c*. 125 m in length and up to 15 m in height. Detailed examination of the constituent facies and the identification of distinct boundaries allowed the construction of a structural interpretation, which is shown in Fig. 5. Five key units, comprising a single facies or combination of facies, were identified and are described in this section. The apparent base of units 2, 3 and 4 is determined by the river level. In reality, they are likely to extend below this point. As the majority of the constituent facies (with the exception of unit 1) comprised ice-rich permafrost, the cryofacies classification devised by Murton & French (1994) has been employed to aid their description. The time spent working on the sequence was restricted by frequent rockfalls from the top of the sequence and the associated dangerous working conditions. This limited the level of descriptive detail that could be recorded, concerning the sedimentology and ice contents of the individual units.

Unit 1: ice-free diamicton

The entire section was capped by a layer of sand-rich diamicton up to 2 m in thickness (Fig. 6A). The diamicton contained abundant sub-rounded and sub-angular clasts mostly between 5 and 10 cm, but occasionally up to 2 m in length (*A*-axis). The unit displayed a clear sub-horizontal

Eastern (proximal) end

Ice-flow direction ⟶

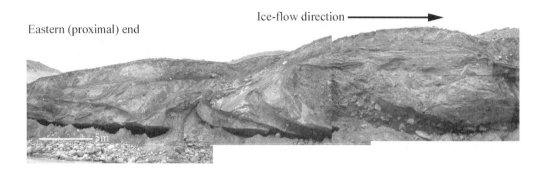

Western (distal) end

Fig. 4. Photo-montage of the moraine section. The montage is split to aid clarity. The western end of the upper montage overlaps with the eastern end of the lower montage

stratification, highlighted by occasional layers of silty-sand and a strongly preferred sub-horizontal orientation to the clasts. The unit was also ice-free and unconsolidated as a result. This diamicton was separated from the underlying ice-rich sediments by an angular unconformity, defined by the truncation of structures within the subjacent units, and an abrupt change in ice content.

Unit 2: steeply dipping, ice-rich diamicton

The stoss-side of the section was characterized by an ice-rich diamicton containing layers and

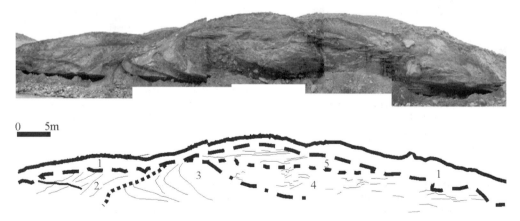

Fig. 5. Structural interpretation of the moraine section. The thick dashed lines indicate major unit boundaries, with different styles of dashing highlighting boundaries between different units. The thin black lines highlight the orientation and extent of structures within the constituent facies

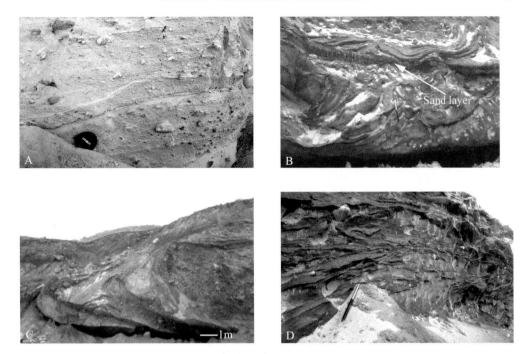

Fig. 6. (a) Unit 1 (lens cap for scale); (b) layer of cross-bedded sands within unit 2 (ice axe for scale); (c) unit 3 (note scale bar); (d) unit 4; note the recumbent, isoclinal folds (ice-axe for scale)

lenses of pure ice and aggregate-poor ice up to 50 cm thick. The ice-rich diamicton was massive, with a matrix dominated by fine- to medium-grained sand and a minor component of silt. The unit contained abundant clasts dispersed throughout the matrix, ranging from granule gravel to clasts up to 50 cm in length (A-axis). These sediments were cemented by pore ice, and excavation and melting of three samples of the ice-rich diamicton revealed a volumetric ice content close to 50%. The pure ice was bubble-rich and sediment-free, whilst the aggregate-poor ice was bubble-poor and characterized by dispersed, angular 'clots' of fine-grained sediment up to 0.5 cm in diameter. The layers and lenses of pure ice and aggregate-poor ice generally dipped up-glacier at c. 40° and were truncated along the overlying unconformity with unit 1. In one location a thin layer of pure ice (10 cm thick) within ice-rich diamicton sandwiched in between thick units of pure ice and aggregate-poor ice (>2 m thick), was deformed into a tight fold with a wavelength of c. 30 cm and an amplitude of c. 15 cm. The sequence also featured a sub-horizontal layer of well-sorted, cross-bedded fine-, medium- and coarse-grained sands up to 50 cm thick, which cross-cut the ice layers within the ice-rich

diamicton (Fig. 6B). As with the ice-rich diamicton, these sands were cemented by pore ice. This layer extended for 10 m across the stoss-side of the unit, before pinching out towards the west.

Unit 3: folded, layered sequence

The central part of the moraine section was characterized by a layered sequence comprising intercalated layers of ice-rich diamicton, aggregate-poor ice and ice-rich gravel between c. 1 and 6 m thick (Fig. 6C). The ice-rich diamicton and aggregate-poor ice were identical in character to those described in the previous section. The ice-rich gravel contained a dense concentration of sub-rounded clasts up to c. 1 m in length (A-axis) set within a matrix of medium- to coarse-grained sand. This facies was also massive in appearance, lacking any clear internal structure. As with the ice-rich diamicton, the ice-poor gravel was cemented by pore ice although no excess ice was evident, suggesting an ice content of 25–50% by volume. The entire layered sequence was deformed into a fold with a near-horizontal axial surface.

Units 2 and 3 were separated by a major discontinuity that dipped towards the east. The discontinuity was associated with a throw of at least the thickness of the exposed sequence and is interpreted as a thrust fault. The fault also contained a lens of bubble-rich ice *c.* 1.5 m long and 0.5 m thick.

Unit 4: thinly layered ice-rich diamicton

Unit 3 graded laterally into unit 4, which dominated the distal part of the moraine section. The unit was very similar to unit 2 both in terms of the texture of its constituent sediments and the occurrence of thin ($\leq$30 cm thick), laterally discontinuous layers and lenses of pure ice and aggregate-poor ice (Fig. 6D). In contrast to the regular dip of the majority of these layers and lenses in unit 2, in unit 4 they were deformed into a series of recumbent, isoclinal folds, with strongly attenuated limbs (Fig. 6D). These folds were characterized by wavelengths between 20 and 60 cm, although their amplitudes were difficult to determine due to the pinching-out of the fold limbs.

Unit 5: layered upper unit of ice-rich diamicton

Examination of this unit was precluded by its inaccessibility. Consequently, the description of the facies and bounding contacts provided below is based upon observation from the other side of the river.

Unit 5 appeared to be composed primarily of ice-rich diamicton with sub-horizontal layers and lenses of debris-poor ice. It is likely that the sedimentology and ice content (laterally discontinuous layers and lenses of aggregate-poor ice and pure ice) were similar to units 2 and 4, although this was impossible to confirm in the field. Unit 5 was underlain by a sub-horizontal, planar unconformity where it was situated above unit 3, indicated by a clear contrast in sedimentology. Continuation of this same unconformity was visible to the west (down-glacier), where unit 5 overlay unit 4, delineated by a contrast in the characteristics of the included ice layers and lenses (sub-horizontal and planar in unit 5; inclined and deformed in unit 4). Unit 5 was also separated from unit 1 above by a curved unconformity (concave-downwards), almost certainly associated with an abrupt change in ice content.

Moraine section – interpretation

Origin of the constituent facies

Before interpreting the sequence as a whole, it is first necessary to consider the origin of the constituent facies. The sequence comprises a variety of ice and sediment facies, the most common of which is ice-rich diamicton. This is identical in appearance to stratified facies basal ice exposed at the current ice margin and the nearby Russell Glacier, comprising a poorly sorted, sand-rich diamicton cemented by pore ice. This basal ice facies commonly contains layers and lenses of debris-poor ice that have been deformed to produce recumbent, isoclinal folds similar to those observed within units 2 and 4, and is associated with the basal freeze-on of sediment, water and snow at the cold-based margin (see Knight *et al.* 1994; Waller *et al.* 2000 for detailed descriptions). In addition, the aggregate-poor ice is similar in character to the dispersed facies basal ice that commonly overlies the stratified facies basal ice at the current ice margin, being characterised by dispersed, angular clots of silt (e.g., Knight *et al.* 1994). Finally, the pure ice is similar in appearance to the firnified, englacial ice, particularly in terms of its lack of included debris and relatively high bubble content. These three ice facies are commonly intercalated at the contemporary ice-sheet margin by a combination of folding and thrusting (Knight 1989; Knight *et al.* 1994), associated with ice deceleration and flow-parallel, compressive strain (Knight 1992). Consequently, the units that comprise mixtures of ice-rich diamicton, aggregate-poor ice and pure ice (units 2, 4 and 5) are considered to represent relict bodies of englacial and basal ice preserved within the moraine complex following ice retreat.

Other major facies include the ice-rich gravel in the central part of the section (unit 3) and the ice-free diamicton that caps the whole sequence (unit 1). Being dominated by sand and gravel and an abundance of large, sub-rounded clasts, the ice-rich gravel is attributed to meltwater activity and glaciofluvial deposition on small sandur, similar to those seen in the current foreland. The ice-free diamicton capping the sequence is interpreted as a melt-out till, as it is: (a) underlain by a thaw unconformity; (b) similar in thickness to the local active layer; and (c) stratified in character (Haldorsen & Shaw 1982).

Origin of the moraine complex

Although the section was excellently exposed, it only cut through part of the moraine complex. In addition, it is likely that only the upper part of a thicker sequence was exposed. Therefore, whilst the moraine section represents a glaciotectonically-deformed sequence of ice and frozen sediment, a degree of uncertainty regarding its exact mode of origin is inevitable.

The first potential interpretation is that the sequence, and the moraine complex as a whole, represents a relict basal ice layer that stagnated during retreat of the ice margin, and which has been preserved by the prevailing permafrost regime. Whilst much of the sequence comprises a range of englacial and basal ice facies, an origin associated solely with ice retreat is, however, considered unlikely for a number of reasons. Firstly, the scale and complexity of folding and faulting displayed within the sequence is greater than that encountered within the basal ice layer exposed around the current margin of the Russell and Leverett Glaciers (Knight *et al.* 1994; Waller *et al.* 2000). Secondly, the most conservative structural interpretation of the exposed section requires large compressive strains in order to juxtapose units of contrasting deformation styles, with considerable associated thickening of the sequence. It is difficult to envisage how the compressive strains necessary to produce the observed structures, and to create the major landform of the moraine complex itself, could have been created by a stagnating and retreating ice margin. Thirdly, it is also unclear how a major erosional unconformity, evident within the central part sequence, could have been generated within a basal ice layer.

It is argued that the structural characteristics of the sequence are best explained by successive phases of ice advance and compression, subglacial erosion and ice retreat. Whilst appearing more complex, this interpretation is consistent with the glacier's known history of significant ice marginal oscillations (van Tatenhove 1995). Details of this model are presented below and illustrated in Fig. 7.

Stage 1 – ice retreat

Ice retreat, most likely during the Holocene Hypsithermal, led to stagnation and burial of intercalated blocks of englacial and basal ice, depositing layers and lenses of ice-rich diamicton, aggregate-poor ice and pure ice. Meltwater activity also led to the aggradation of glaciofluvial material and the deposition of layers and lenses of sorted sediment. These were later frozen as permafrost aggraded downwards through the deglaciated sequence. Such is the complexity of glacial depositional environments that the resulting sequence is likely to have displayed significant local-scale facies variations. Figure 7 (stage 1) depicts the geometry of facies required by the following structural

interpretation, with a proximal layer of ice-rich diamicton (unit 2) occurring at a lower level than a more distal layered sequence of ice-rich diamicton, aggregate-poor ice and ice-poor gravel (unit 3), overlain by a second layer of ice-poor diamicton (unit 4).

Stage 2 – ice advance and compression

Ice retreat was followed by a readvance. This may reflect the glaciological response to climatic deterioration following the Holocene Hypsithermal (van Tatenhove 1995). This resulted in longitudinal compression of the sequence, the formation of the thrust fault between units 2 and 3, and the associated drag fold deformation of units 3 and 4 (Fig. 7 – stage 2). The throw must have been at least 7 m (thickness of the sequence). Unit 2 was also tilted as part of the thrust ramp, as indicated by the inclined ice lenses and layers, indicating that it must have been located at a lower level than units 3 and 4 prior to deformation. This compression may have been associated with the proglacial deformation of permafrost, in which case units 2, 3 and 4 represent pre-existing units occurring at different levels within the proglacial stratigraphy (Fig. 7 – stage 1). Alternatively, unit 2 may represent the basal ice layer of the ice margin advancing over a pre-existing landform composed of units 3 and 4, in which case the thrust represents the glacier bed. In this case, the lens of bubble-rich ice is likely to represent snow entrained into the basal ice layer as the ice overrode the ice-proximal slope of the pre-existing landform.

Stage 3 – continued ice advance and subglacial erosion

Continued ice advance led to the site being overridden and becoming subglacial. This interpretation is based upon the presence of the erosional unconformity overlying units 3 and 4, and the associated removal of a substantial thickness of material (Fig. 7 – stage 3). As units 2 and 3 are distinct, a minimum of 7 m of material must have been removed, equivalent to the exposed thickness of unit 3. It is argued that the removal of this quantity of material could only have occurred across a major décollement, and therefore that this planar unconformity represents the former glacier bed or the base of an associated deforming layer.

An alternative model rests upon the relationship between units 2 and 5 (Fig. 7 – model 2). Whilst the first model assumes they are unrelated, the alternative model assumes units 2

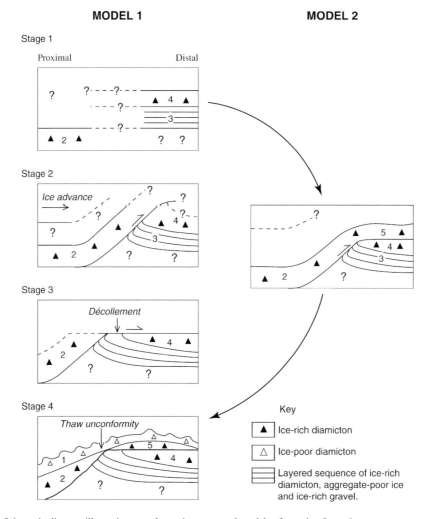

Fig. 7. Schematic diagram illustrating two alternative structural models of moraine formation

and 5 represent the same unit (unfortunately, the continuity or otherwise of units 2 and 5 cannot be established, since they are separated by the unconformity at the base of unit 1). As the ice advanced, unit 2 overrode units 3 and 4 to become unit 5, and the sub-horizontal unconformity overlying units 3 and 4 represents a continuation of the thrust ramp (Fig. 7 – model 2). Continued ice advance and subglacial erosion then resulted in the partial destruction of this drag fold. In this case, unit 2/5 is most likely to represent the basal ice layer of the glacier as it overrode a pre-existing landform composed of units 3 and 4.

Whichever model is assumed to be correct, the removal of material along the erosional

unconformity means the current moraine complex is likely to be much lower than it was prior to it being overridden.

Stage 4 – ice retreat and exposure

The final stage involves ice retreat and the exposure of the moraine complex. van Tatenhove (1995) suggests this is likely to have occurred during the first half of the twentieth century. Ice retreat resulted in the deposition of unit 5 as a relict basal ice layer on top of the erosional décollement surface (Fig. 7 – stage 4). It also resulted in the deposition of a series of small recessional moraine ridges, evident across the entire moraine complex and associated with brief stillstands and melt-out from the basal ice layer. Finally, the thaw of the surface layers following exposure resulted in the production of a melt-out till covering the site (unit 1), and the establishment and progressive development of a series of thaw lakes (Fig. 2).

The origin of the horizontal layer of cross-bedded sands in unit 2 is enigmatic. It clearly post-dates the deformation of unit 2 as it cross-cuts the tilted ice layers. However, its location within a confined space in the mid-part of the section is hard to explain. It may represent the product of thermal erosion and subsequent deposition by meltwater at a time when the base level was higher, although this merely represents an educated hypothesis.

The origin and timing of the small scale folding within units 2 and 4 are unconstrained. It is possible that this deformation is contemporaneous with the formation of the larger structures that define the unit boundaries. The amplitude of the recumbent folds in unit 4 suggest a deformational strain much greater than the minimum throw on the thrust fault that can be estimated stratigraphically. There is, however, insufficient information in the exposed section to constrain the temporal relationship between this folding and the thrust. This presents the possibility that either the folds are contemporaneous, and the thrust structure is part of a regional large-strain compressional regime, or alternatively that unit 4 has undergone a compressional event of sufficient magnitude to generate a tight recumbant fold at a time prior to the development of the thrust structure.

Discussion

The majority of push-moraines associated with Arctic glaciers and glacier–permafrost interactions are believed to be associated with proglacial deformation and the stacking of imbricate thrust sheets. The resulting push-moraine complexes, often termed 'thrust-block moraines' have been described at a number of sites in the Canadian Arctic Archipelago in particular (Evans & England 1991). Lehmann (1992), for example, suggested that advance of the Thompson Glacier, Axel Heiberg Island, is associated with the thrusting, tilting and stacking of slabs of permafrozen glaciofluvial material. As ice advance continues, a basal thrust underlying the whole moraine propagates distally, resulting in the thrusting and incorporation of new blocks. Boulton et al. (1999) related this process explicitly to permafrost, equating the basal thrust with the base of the permafrost layer. Such a mechanism might appear appropriate for the moraine complex at the Leverett Glacier on the

basis of its surface geomorphology. The presence of an elevated zone, characterized by a series of arcuate moraine ridges orientated transverse to ice flow appears consistent with an origin through proglacial deformation, with the moraine ridges representing individual thrust blocks. However, both the internal structure and the mode of origin of the moraine complex at the Leverett Glacier differ markedly from classic thrust-block moraines in two respects: firstly, many of the facies that comprise the complex are ice-marginal or subglacial, rather than proglacial in origin and, secondly, the complex has experienced a significant amount of subglacial erosion. In combination, this indicates a sustained ice advance in which either a pre-existing landform, or a push-moraine relating to the initial stages of the advance, has been overridden and substantially modified as a result.

The predominance of ice-marginal or subglacial facies may reflect the operation of two processes, singularly or in combination. Firstly, as previously mentioned, the occurrence and intercalation of various basal and englacial ice facies may reflect compression and deformation of the ice margin itself (Knight 1989; Knight et al. 1994). However, the inclusion of ice-rich gravel suggests that proglacial materials were also incorporated. Secondly, these ice facies may have been deposited during ice retreat as parts of the ice margin stagnated, subsequently becoming part of the proglacial stratigraphy and being preserved by permafrost aggradation. Later readvance then resulted in these materials being reactivated and reworked.

In terms of its structural characteristics and the role played by subglacial processes, Evans's (1989) description of the nature of glaciotectonic structures at the margins of sub-polar glaciers on Ellesmere Island provides a valuable comparison. At one of his sites (glacier 4), he described a similar situation to that envisaged at the Leverett Glacier, where sustained ice advance has resulted in the overriding of a pre-existing moraine, leading in turn to subglacial shearing and the deformation of the underlying materials. In this case, however, the materials are recycled and incorporated into the basal ice layer of the overriding glacier as frozen rafts of sediment, whereas at the Leverett, subglacial deformation and erosion has resulted in the removal of a substantial thickness material, and the formation of a distinct planar unconformity. The potential impact of subglacial deformation and erosion is highlighted by Hart (1994), who described a Late Weichselian sequence in Iceland resulting from successive phases of proglacial and subglacial deformation. This produced a structural signature similar to that found at the Leverett, with reverse faulting and overturned bedding associated with proglacial compression, and subsequent subglacial deformation, erosion and the production of a sub-horizontal unconformity. In this case, however, the subglacial deformation and associated erosion resulted in the complete destruction of the original landform.

Boulton et al. (1999) provided a set of criteria likely to promote the formation of large, broad push-moraines that potentially explain why the moraine complex at the Leverett Glacier was overridden, in contrast to other Arctic push moraines. One of the key factors thought to allow the distal progradation of push moraines is that the permafrost is broken distally. This is frequently the case in Spitsbergen, for example, where many push moraines occur in emergent coastal locations associated with a tapering wedge of permafrost. In contrast, the Leverett Glacier terminates in a confined basin, where the permafrost layer is unbroken (van Tatenhove, 1995). This situation would have restricted the distal movement of the push-moraine and promoted its overriding. Additionally, the greater thickness of the permafrost and depth of the likely décollement would also limit the mobility of the permafrost layer and thereby encourage overriding. The presence of deformed ice lenses, marked erosional unconformities and truncated structures within the section corroborates recent work suggesting subglacial sediment deformation and erosion can remain active at subfreezing temperatures (e.g., Echelmeyer & Zhongxiang 1987; Fitzsimons 1996; Cuffey et al. 1999, 2000; Fitzsimons et al. 2001; Atkins et al. 2002; Bennett et al. 2003). Such behaviour is promoted by relatively warm temperatures close to the pressure melting, and the associated retention of liquid water adsorbed to particle surfaces, particularly in fine-grained sediments, producing what Tsytovich (1975) termed 'plastic-frozen ground'.

In view of the significant role played by subglacial deformation and erosion, it is debatable whether the final moraine complex represents a push-moraine sensu stricto. Whilst the moraine complex has retained a clear landscape expression, despite being overridden, its structural characteristics have been heavily modified and overprinted by its subsequent overriding. This has resulted in a mismatch between its surface geomorphology, associated with ice retreat and the deposition of small recessional moraines, and its internal structure, reflecting its complex, polygenetic origin. Although the resulting landform can be considered a push-moraine in the broad, non-genetic sense of the term, its description as an ice-marginal glaciotectonic landform is perhaps more accurate. This also highlights the dangers associated with interpreting the genesis of such features on the basis of geomorphology alone.

The description of the moraine complex at the Leverett Glacier adds to both the broad range of glaciotectonic landforms discussed within the literature, and to the range of mechanisms proposed for their formation. It also highlights the influential role played by subglacial processes beneath cold-based ice. Unfortunately, the significant ice content of the complex as a whole suggests that the preservation potential of the diagnostic structures is limited, unless the permafrost regime prevails and the constituent units are preserved beneath a carapace of melt-out till.

Conclusions

The proglacial area of the Leverett Glacier is dominated by a large, arcuate moraine complex rising 15–20 m above the adjacent sandur surfaces. Whilst its general appearance and surface morphology suggest it represents a multi-crested push moraine, related solely to proglacial deformation, analysis of a stream-cut section reveals a more complex, polygenetic origin. Structural interpretation of the section, composed of a variety of ice and frozen sediment facies, indicates successive phases of both proglacial and subglacial deformation. Initial ice advance, possibly following the Holocene Hypsithermal, was accompanied by proglacial deformation, and the development of a thrust fault and associated drag fold. The advancing ice subsequently overrode the site, resulting in significant subglacial deformation and erosion, and the truncation of pre-existing structures along a pronounced planar discontinuity. Finally, ice retreat led to the deposition of a series of small recessional moraine ridges on the surface of the complex, whilst ablation led to the thaw of a capping layer of melt-out till and the creation of a series of thaw lakes.

The moraine's genesis through a combination of proglacial and subglacial deformation contrasts with the majority of multi-crested push moraines in permafrost regions, which are primarily related solely to the proglacial thrusting of frozen sheets of sediment. As such, the resulting polygenetic feature is more accurately described as an ice-marginal, glaciotectonic landform. The occurrence of overriding as opposed to proglacial deformation may reflect the influence of a thick and spatially continuous layer of permafrost on the rheology of the foreland, producing a landform that adds to the variety of previously published features associated with glacier–permafrost interactions.

Financial support enabling field investigations in 2000 and 2002 was provided by the School of Earth and Environmental Sciences, University of Greenwich, and the School of Earth Sciences and Geography, Keele University, respectively. Logistical support and research permits provided by the Danish Polar Centre and the Kangerlussuaq Institute for Scientific Support are gratefully acknowledged. The authors would like to thank Will Adam and Zoe Robinson for invaluable field assistance and Andy Lawrence for cartographic support. Finally, the authors would also like to extend their gratitude to Professor Geoffrey Boulton, whose insightful questioning led to the formulation of the alternative model, and to Professor Matthew Bennett and Dr. Sean Fitzsimons, whose constructive reviews led to substantial improvements to this paper.

References

ASTAKHOV, V. I., KAPLYANSKAYA, F. A. & TARNOGRADSKIY, V. D. 1996. Pleistocene permafrost of West Siberia as a deformable glacier bed. *Permafrost and Periglacial Processes*, **7**, 165–191.

ATKINS, C. B., BARRETT, P. J. & HICOCK, S. R. 2002. Cold glaciers erode and deposit: evidence from Allan Hills, Antarctica. *Geology*, **30**, 659–662.

BENNETT, M. R. 2001. The morphology, structural evolution and significance of push moraines. *Earth-Science Reviews*, **53**, 197–236.

BENNETT, M. R., WALLER, R. I., MIDGLEY, N. G., HUDDART, D., GONZALEZ, S., COOK, S. J. & TOMIO, A. 2003. Subglacial deformation at sub-freezing temperatures? Evidence from Hagafellsjökull-Eystri, Iceland. *Quaternary Science Reviews*, **22**, 915–923.

BOULTON, G. S., VAN DER MEER, J. J. M., BEETS, D. J., HART, J. K. & RUEGG, G. H. J. 1999. The sedimentary and structural evolution of a recent push moraine complex: Holmstrømbreen, Spitsbergen. *Quaternary Science Reviews*, **18**, 339–371.

CLAYTON, L., ATTIG, J. W. & MICKELSON, D. M. 2001. Effects of late Pleistocene permafrost on the landscape of Wisconsin, U.S.A. *Boreas*, **30**, 173–188.

CUFFEY, K. M., CONWAY, H., HALLET, B., GADES, A. M. & RAYMOND, C. F. 1999. Interfacial water in polar glaciers, and glacier sliding at $-17°C$. *Geophysical Research Letters*, **26**, 751–754.

CUFFEY, K. M., CONWAY, H., GADES, A. M., HALLET, B., LORRAIN, R., SEVERINGHAUS, J. P., STEIG, E. J., VAUGHN, B. & WHITE, J. W. C. 2000. Entrainment at cold glacier beds. *Geology*, **28**, 351–354.

CUTLER, P. M., MACAYEAL, D. R., MICKELSON, D. M., PARIZEK, B. R. & COLGAN, P. M. 2000. A numerical investigation of ice-lobe–permafrost interaction around the southern Laurentide ice sheet. *Journal of Glaciology*, **46**, 311–325.

DOWDESWELL, J. A. & SIEGERT, M. J. 1999. Ice-sheet numerical modelling and marine geophysical measurement of glacier-derived sedimentation on the Eurasian Arctic continental margins. *Geological Society of America Bulletin*, **111**, 1080–1097.

ECHELMEYER, K. & ZHONGXIANG, W. 1987. Direct observation of basal sliding and deformation of basal drift at subfreezing temperatures. *Journal of Glaciology*, **33**, 83–98.

ETZELMÜLLER, B., HAGEN, J. O., VATNE, G., ØDEGÅRD, R. S. & SOLLID, J. L. 1996. Glacier debris accumulation and sediment deformation influence by permafrost: examples from Svalbard. *Annals of Glaciology*, **22**, 53–62.

EVANS, D. J. A. 1989. The nature of glacitectonic structures and sediments at sub-polar glacier margins, northwest Ellesmere Island, Canada. *Geografiska Annaler*, **71A**, 113–123.

EVANS, D. J. A. & ENGLAND, J. 1991. Canadian landform examples 19, High Arctic thrust block moraines. *Canadian Geographer*, **35**, 93–97.

FITZSIMONS, S. J. 1996. Formation of thrust-block moraines at the margins of dry-based glaciers, south Victoria Land, Antarctica. *Annals of Glaciology*, **22**, 68–74.

FITZSIMONS, S. J., MCMANUS, K. J., SIROTA, P. & LORRAIN, R. D. 2001. Direct shear tests of materials from a cold glacier: implications for landform development. *Quaternary International*, **86**, 129–137.

HALDORSEN, S. & SHAW, J. 1982. The problem of recognising melt-out till. *Boreas*, **11**, 261–277.

HART, J. K. 1994. Proglacial glaciotectonic deformation at Melabakkar-Ásbakkar, west Iceland. *Boreas*, **23**, 112–121.

HART, J. K. & WATTS, R. J. 1997. A comparison of the styles of deformation associated with two recent push moraines, south van Keulenfjorden, Svalbard. *Earth Surface Processes and Landforms*, **22**, 1089–1107.

HUDDART, D. & HAMBREY, M. J. 1996. Sedimentary and tectonic development of a High-Arctic, thrust-moraine complex: Comfortlessbreen, Svalbard. *Boreas*, **25**, 227–243.

KÄLIN, M. 1971. The active push moraine of the Thompson Glacier. Axel Heiberg Island Research Reports, Glaciology no. 4. McGill University, Montreal.

KLEMAN, J. 1994. Preservation of landforms under ice sheets and ice caps. *Geomorphology*, **9**, 19–32.

KNIGHT, P. G. 1989. Stacking of basal debris layers without bulk freezing-on: isotopic evidence from West Greenland. *Journal of Glaciology*, **35**, 214–216.

KNIGHT, P. G. 1992. Ice deformation very close to the ice-sheet margin in West Greenland. *Journal of Glaciology*, **38**, 3–8.

KNIGHT, P. G., SUGDEN, D. E. & MINTY, C. D. 1994. Ice flow around large obstacles as indicated by basal ice exposed at the margin of the Greenland ice sheet. *Journal of Glaciology*, **40**, 359–367.

KRÜGER, J. 1996. Moraine ridges formed from subglacial frozen-on sediment slabs and their differentiation from push moraines. *Boreas*, **25**, 57–63.

LEHMANN, R. 1992. Arctic push moraines, a case study
     of the Thompson Glacier Moraine, Axel Heiberg
     Island, N.W.T., Canada. *Zeitschrift für Geomor-
     phologie N.F.*, **86** (suppl.), 161–171.
MACKAY, J. R. 1959. Glacier ice – thrust features of
     the Yukon Coast. *Geographical Bulletin of
     Canada*, **13**, 5–21.
MOOERS, H. D. 1990. Ice-marginal thrusting of drift
     and bedrock: thermal regime, subglacial aquifers,
     and glacial surges. *Canadian Journal of Earth
     Sciences*, **27**, 849–862.
MORAN, S. R., CLAYTON, L., HOOKE, R. L.,
     FENTON, M. M. & ANDRIASHEK, L. D. 1980.
     Glacier-bed landforms of the Prairie Region of
     North America. *Journal of Glaciology*, **25**, 457–476.
MURTON, J. B. & FRENCH, H. M. 1994. Cryostructures
     in permafrost, Tuktoyaktuk Coastlands, Western
     Arctic Canada. *Canadian Journal of Earth
     Sciences*, **31**, 737–747.
MURTON, J. B., WALLER, R. I., HART, J. K.,
     WHITEMAN, C. A., POLLARD, W. & CLARK, I.
     2005. Stratigraphy and glaciotectonic structures
     of permafrost deformed beneath the northwest
     margin of the Laurentide Ice Sheet, Tuktoyaktuk
     Coastlands. *Journal of Glaciology*, **170** (in press).
PATERSON, W. S. B. 1994. *The Physics of Glaciers*, 3rd
     edn Pergamon Press, Oxford.
SCHOLZ, H. & BAUMANN, M. 1997. An 'open system
     pingo' near Kangerlussuaq (Søndre Strømfjord),
     West Greenland. *Geology of Greenland Survey
     Bulletin*, **176**, 104–108.
TSYTOVICH, N. A. 1975. *The Mechanics of Frozen
     Ground*. McGraw-Hill, New York.
VAN TATENHOVE, F. G. M. 1995. The dynamics of
     Holocene deglaciation in west Greenland with
     emphasis on recent ice-marginal processes.
     Universiteit van Amsterdam, Amsterdam.
WALLER, R. I., HART, J. K. & KNIGHT, P. G. 2000. The
     influence of tectonic deformation on facies varia-
     bility in stratified debris-rich basal ice. *Quaternary
     Science Reviews*, **19**, 775–786.

The interaction of a surging glacier with a seasonally frozen foreland: Hagafellsjökull-Eystri, Iceland

MATTHEW R. BENNETT[1], DAVID HUDDART[2] & RICHARD I. WALLER[3]

[1]*School of Conservation Sciences, Bournemouth University, Talbot Campus, Poole, Dorset BH12 5BB, UK*
(e-mail: mbennett@bournemouth.ac.uk)
[2]*School of Outdoor Education, Liverpool John Moores University, I.M. Marsh Campus, Barkhill Road, Liverpool L19 3DB, UK*
[3]*School of Earth Sciences and Geography, Keele University, Staffordshire ST5 5BG, UK*

Abstract: This paper describes aspects of the landform–sediment assemblage produced by a recent surge of Hagafellsjökull-Eystri. This surge occurred during the winter and early spring of 1998/1999 and consequently advanced into a partially frozen foreland. Two aspects of this landform–sediment assemblage are considered. First the evidence for a frozen subglacial sediment layer beneath a lateral piedmont lobe formed during the surge is reviewed. This sediment layer consists of blocks of glacier ice set within a matrix of frozen sediment and was injected into basal crevasses to form a network of crevasse-squeeze ridges prior to freezing. The sediment layer appears to provide evidence of sub-freezing deformation at the termination of the surge. Secondly the paper examines the detailed tectonic facies within a push-moraine formed along the eastern latero-frontal margin of the glacier during the surge. Architecturally this push moraine consists of a multi-layered slab of glaciofluvial sediments with a monocline structure that has been displaced laterally by the advancing ice margin. The sediment slabs within this monocline are characterized by both brittle and ductile styles of deformation. The authors argue that the observed variation in deformation style may be explained by spatial variation in the extent to which the glacial foreland was frozen or unfrozen at the time of displacement. Areas of frozen foreland would have behaved in a brittle fashion, while unfrozen areas deformed in a more ductile manor. Both these aspects of the landform–sediment assemblage examined in this paper appear to be the product of the seasonal timing of the surge. Not only do they add to our understanding of surge-type landsystems, but they also illustrate the potential of winter advances around the margins of some temperate glaciers to explore the coupling between glaciers and frozen proglacial sediments.

The sediment–landform assemblage formed by glacier surges has been well documented over the last 20 years from a variety of geographical locations and from both terrestrial and marine settings (Sharp 1988; Solheim 1991; Boulton *et al.* 1996; Evans & Rea 1999). This assemblage is characterized by the formation of a glaciotectonic moraine system along the surge limit, and by a network of crevasse-squeeze ridges inside these limits (Sharp 1985a, b). Crevasse-squeeze ridges form as a result of the injection of subglacial sediment into fractured basal ice at the end of the surge as the glacier become quiescent and settles into its saturated bed (Sharp 1985a). Rapid glaciofluvial sedimentation is often associated with this final phase during the ablation of the post-surge glacier tongue. This landsystem is considered to be diagnostic of surge-type behaviour within the geological record (Sharp 1988; Evans & Rea 1999). Caution is, however, required given that the preservation potential of such features as crevasse-squeeze ridges is poor and the formation of glaciotectonic moraine systems is not necessarily restricted to surge-type glaciers (Hambrey & Huddart 1995). It is also important to note that, although this landform assemblage appears to be fairly consistent across a range of different geographical locations, the influence of different environmental characteristics have yet to be fully factored into this landsystem model. The current paper contributes to our understanding of this

From: HARRIS, C. & MURTON, J. B. (eds) 2005. *Cryospheric Systems: Glaciers and Permafrost.*
Geological Society, London, Special Publications, **242**, 51–62.
0305-8719/05/$15.00 © The Geological Society of London 2005.

landsystem by examining the sediment–land-form assemblage associated with a glacier that surged during the winter–spring months, and consequently advanced into a seasonally frozen foreland. We focus on Hagafellsjökull-Eystri an outlet glacier of the Langjökull ice cap in central Iceland (Fig. 1). Bennett *et al.* (2000) documented the glacial foreland prior to the surge, while this paper provides an overview of the post-surge sediment–landform assemblage.

Hagafellsjökull-Eystri

Hagafellsjökull-Eystri is a south-flowing, 4 km-wide, outlet glacier of the Langjökull ice cap

(Figs 1 and 3). It currently terminates in a progla-cial lake (Hagavatn) and is constrained laterally to the east by the volcanic ridges of the Jarlhettur. The glacier overflows cols in the Jarlhettur to form a series of piedmont lobes in the adjacent Jarlhettukvisl Valley. The glacier surged in 1975, 1980 and 1999 (Fig. 2; Sigbjarnarson 1975; Theódórsson 1980; Sigurðsson 2001). No surges are known before 1975, although the land-form assemblage that marks the 1890 Neoglacial ice maximum may have been associated with several surge-like ice-marginal oscillations. During each of the three recent surges, the glacier advanced between 1000 and 1500 m during a relatively short period of late winter or

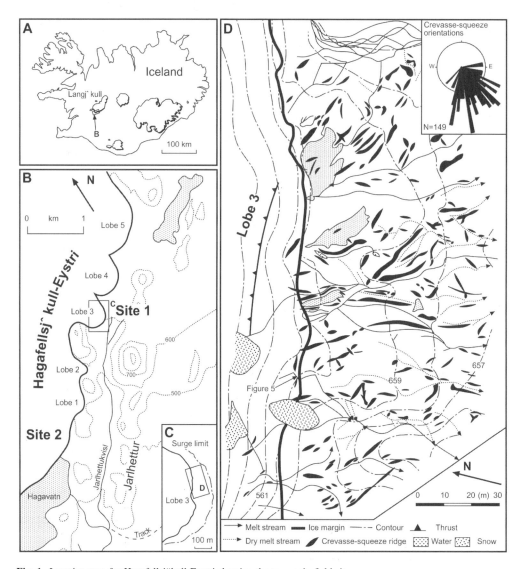

Fig. 1. Location map for Hagafellsjökull-Eystri showing the two main field sites

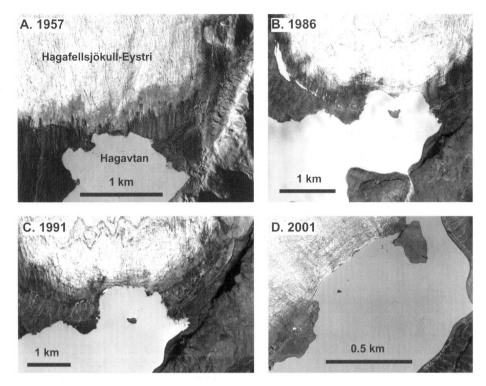

Fig. 2. Vertical aerial photographs of Hagafellsjökull-Eystri at various times since 1957. Note the presence of two subaqueous fans after the surges of 1980 and 1998/1999. A–C reproduced with courtesy of Land Maelingar Island. D reproduced with courtesy of NERC

early spring. The most recent surge in 1998/1999 was no exception. The surge started in the autumn of 1998 when the lateral piedmont lobes in the Jarlhettukvisl Valley began to advance. The frontal terminus of the glacier did not, however, advance until April 1999, when it moved over 30 m in the first 24 h. The glacier front advanced a total of 1165 m over the following six weeks at a rate of 25 m per day (Sigurðsson 2001; Fig. 3A, B). The authors were fortunate to examine the glacier and its foreland prior to the surge in the summer of 1998 and have traced its post-surge evolution during the summers of 2000, 2001, 2002 and 2003. Two areas of the glacier's foreland are considered in this paper, the first is the lateral piedmont lobes in the Jarlhettukvisl Valley and the second is the eastern latero-frontal margin of the glacier (Fig. 1).

Lateral piedmont lobes of the Jarlhettukvisl Valley

Four lateral piedmont lobes formed in the Jarlhettukvisl Valley during the surge of 1998/1999. At the height of the surge these lobes were heavily crevassed. By 2003, however,

surface ablation had produced a relatively crevasse-free ice margin. Lateral retreat of 30–50 m has occurred since the summer of 2001 and has revealed a series of moraine ridges composed of boulder-rich diamicton 2–4 m high, marking the outer limit of the surge. Inside this outer moraine, there is a zone of irregular mounds and enclosed hollows that form a complex and variable local relief that is being rapidly infilled with sediment from numerous small melt-streams. During the summers of 2001 and 2002, the key landform within this zone was a series of rectilinear crevasse-squeeze ridges (Figs 1 and 4) composed of fine-grained diamicton. Individual ridges vary from <2 to >30 m in length and are up to 3 m high and can be traced from the forefield onto the glacier surface. On the glacier surface they consist of vertical dykes of diamicton that emerge from fractures and crevasses, which widen towards the glacier bed. By the summer of 2003 these crevasse-squeeze ridges had begun to degrade, and the forefield was dominated by a low undulating topography consisting of re-worked crevasse-squeeze ridges and small outwash fans.

This sediment–landform assemblage is similar to that documented from surge-type

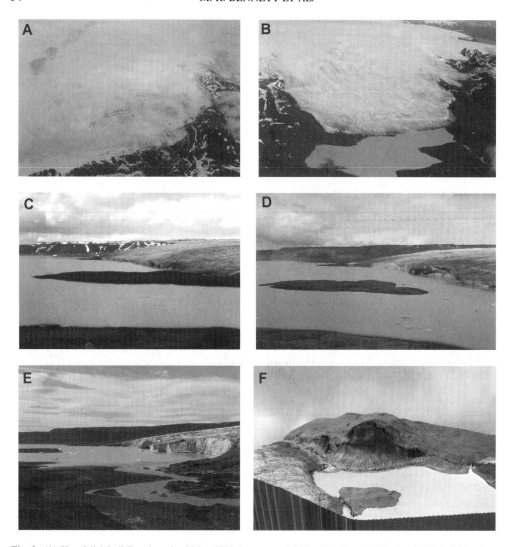

Fig. 3. (**A**) Hagafellsjökull-Eystri on the 6 May 1999 (courtesy of Oddur Sigurðsson). (**B**) Hagafellsjökull-Eystri on the 20 June 1999 (courtesy of Oddur Sigurðsson). (**C**) The margin of Hagafellsjökull-Eystri, showing the emergent subaqueous fans, in June 2001. (**D**) The margin of Hagafellsjökull-Eystri in June 2002. (**E**) The margin of Hagafellsjökull-Eystri in June 2003. (**F**) Digital elevation model generated from aerial photographs of the eastern side of Hagafellsjökull-Eystri. The image shows one of the emergent subaqueous fans during early June 2001

glaciers elsewhere in Iceland (Sharp 1988; Evans & Rea 1999). One key difference, however, was noted during the summer of 2002 in a series of stream cut sections that transect the ice margin and foreland of one of the lateral piedmont lobes (lobe 3, Fig. 1). These sections have been documented in detail elsewhere and have been interpreted as a frozen subglacial layer (Bennett *et al.* 2003). As illustrated in Fig. 5, this frozen layer consists of sub-rounded, imbricated blocks of glacier ice set within a matrix of frozen diamicton. The ice blocks have deformed and attenuated into the frozen diamicton that contains numerous ice-rich fractures. This sub-glacial layer with its allochthonous ice blocks has in some cases been intruded, prior to freezing, into the crevasse-squeeze ridges of the forefield.

Bennett *et al.* (2003) interpreted these sections as evidence of subglacial deformation at sub-freezing temperatures during the final phase of the surge. They argued that the blocks of

Fig. 4. (**A**) Crevasse-squeeze ridge (lobe 3 Fig. 1) within a lateral piedmont lobe of Hagafellsjökull-Eystri during the summer of 2001. (**B**) Frozen diamicton intruded into a basal crevasse beneath a lateral piedmont lobe of Hagafellsjökull during the summer of 2001

glacier ice were derived from the unstable seracs that characterized the piedmont lobes during the surge and were overridden and entrained along with proglacial sediment as the ice advanced. This proglacial sediment is likely to have consisted of washed subglacial diamicton, which outcrops extensively beyond the surge-limit. The preservation of these ice blocks may have been favoured by the winter/spring timing of the surge. According to Bennett *et al.* (2003), the ice blocks were initially subject to deformation within a wet deforming layer, becoming thermally and mechanically abraded. In addition, the A–B axial plane of the blocks was orientated in the direction of shear and, in some cases, the blocks were stacked to give imbricate clusters. Subsequently the matrix of diamicton appears to have begun to freeze, and consequently to behave in a more brittle fashion as its shear strength increased. At this point the ice blocks became the more ductile component within the sediment layer, being deformed and attenuated along fractures within the frozen matrix (Fig. 5). Bennett *et al.* (2003) suggested, therefore, that

the rheological contrast between the blocks of glacier ice and the sediment changed in response to falling temperature at the termination of the surge.

The cause of this subglacial cooling is uncertain. Melting of the ice blocks would have consumed latent heat, thereby causing a fall in ambient temperature. In addition, heat-flow would have occurred from the deforming layer towards the overridden substratum, if the foreland was frozen as the ice advanced. This is likely given that the advance occurred in the winter–spring of 1998/1999. One might also expect the winter cold wave to have penetrated through the heavily crevassed and relatively thin ice lobe formed by the surge. This cooling trend would have been countered by the frictional heat generated within the deforming sediment and by any basal sliding beneath the ice lobe. This source of heat would, however, have declined as the surge terminated and the ice advance slowed. Alternatively, the initial cooling could have increased the strength of the deforming layer and therefore its resistance to

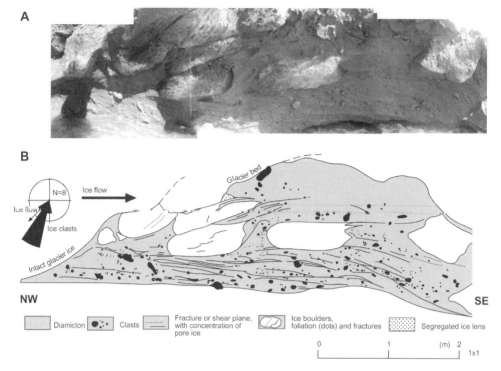

Fig. 5. A section through the frozen subglacial layer beneath lobe 3 of Hagafellsjökull-Eystri in the Jarlhettukvisl Valley. The layer consists of sub-rounded allochthonous blocks of glacier ice set within frozen diamicton

deformation, which in turn would, if sufficient, have slowed the ice advance, thereby reinforcing the cooling trend.

Bennett *et al.* (2003) suggested, therefore, that the timing of the surge and the presence of a seasonally frozen foreland was critical to the formation of this unusual deforming layer. The presence of this frozen, ice-rich layer may also explain the scale of sediment re-working within the glacier foreland and, consequently, the rapid degradation of the crevasse-squeeze ridges.

Eastern latero-frontal margin of Hagafellsjökull-Eystri

Prior to the 1998/1999 surge, the terminus of Hagafellsjökull-Eystri was separated from Hagavatn by between 200 and 1000 m of fluted proglacial foreland (Fig. 2; Bennett *et al.* 2000). The eastern lateral margin was partially detached from the adjacent western ridge of the Jarlhettur by a broad (1 km wide) outwash surface with numerous shallow, braided melt-streams (Fig. 2). Ice-marginal moraines were absent and the ice margin was both low-angled

(*c.* 5°) and crevasse-free. The most distinctive feature of the foreland was the near-surface water-table, giving areas of locally saturated sediment.

By the end of the surge, the glacier margin had re-entered Hagavatn and the eastern margin was within 200 m of the western ridge of the Jarlhettur (Figs 3A, B and 6). During the spring of 1999, as the ice margin stabilized at its maximum position, two large subaqueous outwash fans emerged along the ice margin within Hagavatn (Figs 2D, E, 3C, D, F and 6). These two outwash fans ceased to be active at the end of the surge in July 1999. No data on the subaqueous evolution of either of these fans, or of their relationship to any subaqueous moraine bank, are available. Similar fans formed during earlier surges of the glacier (Fig. 2B, C) and in all cases the fans were abandoned following the surge. The large melt water discharge does not appear to be a feature of the quiescent phases or post-surge glacier. In fact, prior to the 1998/1999 surge, water discharge was confined to two lateral meltwater streams and to groundwater flow. Following the surge the ice-margin calved steadily into Hagavatn at a rate of between 50 and 100 m per annum perpendicular to the ice margin (Fig. 3C–E). The margin became detached from the subaqueous fans during the summer of 2001, and they now form islands within Hagavatn. In 2003, the western side of the ice margin had withdrawn from Hagavatn completely, although the eastern portion of the ice front continues to calve into the lake.

The eastern latero-frontal margin of the surge is marked by a low push-moraine that varies in height from 2 to 5 m and is separated from the western ridge of the Jarlhettur by a relict meltwater channel (Figs 2D and 6). The moraine is composed of combinations of: parallel laminated fine sand, with silt interbeds; alternating units of medium sand and granule gravel; well-sorted granule gravel; poorly sorted, matrix-rich granule gravel with silt stringers; massive, coarse sand units with minor amounts of granule gravel; and parallel laminated silt and fine sand. These sediments are consistent with deposition in a lateral outwash fan subject to a relatively low but variable flow regime, a depositional environment that was typical of the eastern lateral margin of Hagafellsjökull-Eystri prior to the surge (Bennett *et al.* 2000). The gross architecture of the push-moraine consists of a stratigraphically conformable series of sediment layers that rests unconformably on a core of pre-1998, tectonized sediments that form the valley-side drift sheet (Fig. 6). These pre-1998 sediments appear to have acted as a basal décollement surface over which the 1998/1999 push-moraine moved. The over-riding sediment pile that forms the moraine has a monoclinic form, with a fold axis that is parallel to the ridge crest. The crest and distal face of the moraine consists of sub-horizontal sediment layers, while the proximal face is composed of

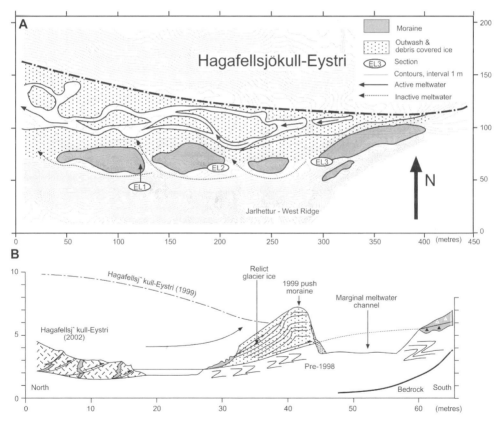

Fig. 6. The push moraine along the eastern lateral margin of Hagafellsjökull-Eystri. (**A**) Contoured geomorphological map of the push-moraine showing the location of excavated sections. The map was generated using a Leica 530 GPS using a real-time kinematic survey. (**B**) Schematic cross-section of the push-moraine formed by the 1998/1999 surge of Hagafellsjökull-Eystri

steeply inclined, or sub-vertical sediment layers, with occasional slivers of buried glacier ice on the ice-contact face (Fig. 6). Along the strike, this monocline shows low-amplitude, long-wavelength undulations, consistent with gentle compression parallel to its crest.

The sediments within the monoclinic structure show contrasting styles of internal deformation. In some locations the transverse, bed-parallel strain was accommodated by brittle deformation, while in other, adjacent locations deformation occurred in a ductile fashion. Three transverse sections dug through the moraine crest illustrate this variation in tectonic style along the length of the moraine (Fig. 6).

(1)   Section EL1 (Fig. 7) exposes two monoclinic, sediment slabs that rest unconformably on a basal décollement composed of fissile, silty clay. Within the upper part of the section, individual units of sediment have been inter-mixed via a series of irregular, isoclinal folds, associated with the ductile deformation of the fine-grained facies around pods of coarse sand and granule gravel under a bedding-parallel strain regime. The tectonic facies is consistent with unfrozen, wet, ductile deformation.

(2)   Section EL2 (Fig. 8) has a similar form to Section EL1 but, in contrast to Section EL1,

the individual sedimentary units within the section have not been intermixed by ductile deformation. Instead, strain has been accommodated via numerous brittle fractures. This is particularly apparent within the horizontally stratified granule-gravel units, several of which contain rotated, fault-bound blocks within a gravel matrix. Finer-grained units within the main body of the moraine show little sign of deformation, despite alternating grain sizes. In addition, the sediment within this moraine is frozen, with abundant interstitial ice, below a depth of 1 m of the moraine surface. The ice-contact face of the moraine also contains a thin sliver of glacier ice, buried beneath a 0.5 m thick layer of slumped sand and gravel.

(3)   Section EL3 (Fig. 9) is again architecturally similar to the previous sections, consisting of a monoclinic structure with a fold axis that is parallel to the ridge axis. As in Section EL2, the strain was accommodated by brittle rather than ductile deformation. For example, the main sand sheet within the centre of the section contains numerous conjugate fractures and faults that show between 1 and 10 cm displacements. Above this sand sheet there is a unit of laminated silty clay. Below the ridge

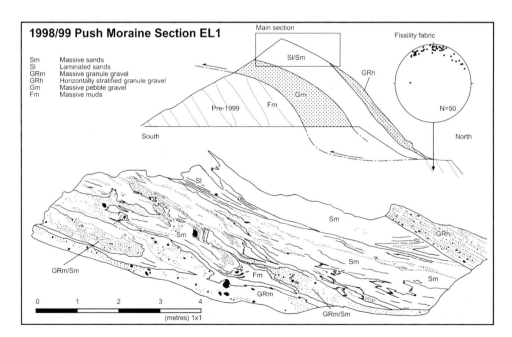

Fig. 7. Field sketch of Section EL1 through the crest of a push-moraine along the eastern lateral margin of Hagafellsjökull-Eystri

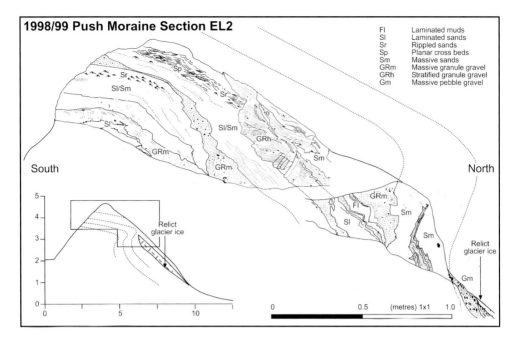

Fig. 8. Field sketch of Section EL2 through the crest of a push-moraine along the eastern lateral margin of Hagafellsjökull-Eystri

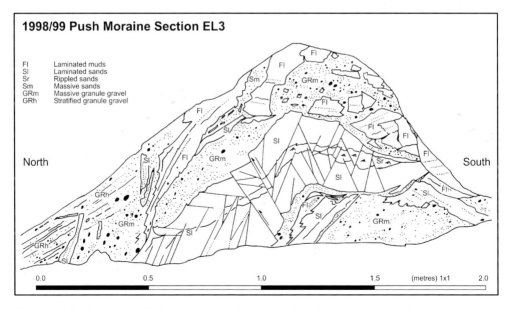

Fig. 9. Field sketch of Section E3 through the crest of a push-moraine along the eastern lateral margin of Hagafellsjökull-Eystri

crest, this unit of silt disintegrates into a series of partially tessellated blocks. As one moves towards the ice-distal face of the moraine, these blocks become increasingly discrete, and are set within a matrix of sand and granule-gravel. In fact, the ice-distal face of the moraine is dominated by a breccia of sediment blocks formed by the disaggregation of sediment sheets at the apex of the moraine crest. The section is dominated, therefore, by brittle deformation and the lee-side disaggregation of the sediment sheets along fractures rather than grain boundaries.

Along the crest of the moraine, the pattern of brittle and ductile deformation, illustrated by Sections EL1–EL3, shows no consistent pattern. Rather, each style is mutually exclusive, despite their close juxtaposition. Explaining this rapid alternation between ductile and brittle styles of deformation along the length of the moraine is challenging. One option is that it may reflect variations in sediment strength during deformation as a result of varying degrees of foreland freezing. Areas of frozen sediment, with relatively small amounts of interstitial ice, are likely to have displayed greater internal cohesion and strength than areas of unfrozen sediment, and will consequently have behaved in a brittle fashion under compressive stress (Broster & Clague 1987). In contrast, unfrozen sections, especially where adjacent areas of frozen terrain impede drainage, leading to high pore-water pressures, are more likely to have undergone ductile deformation when stressed.

Spatial variations in the depth to which a glacier foreland freezes is to be expected, where variations in sediment-water content, and in the distribution and thickness of ice-marginal snow patches occur. Strong katabatic winds from the Langjökull ice cap are likely to have resulted in drifting of significant amounts of snow around the ice-margin, leading to spatial variation in the depth of snow cover. The subsequent impact of snow cover on the ground thermal regime and depth of freezing depends upon the timing as well as the depth of snowfall (Williams & Smith 1989). If snowfall occurs early in the winter, then the underlying ground is insulated from the lowest winter temperatures, and the depth of freezing is subsequently limited. Alternatively, if snowfall occurs later in the season, then existing frozen ground is insulated from rising temperatures and will remain intact for longer into the spring or summer. In addition, a high water content will limit the depth of seasonal freezing, as available pore-water must freeze before the freezing front can penetrate deeper into the substrate. The authors suggest, therefore, that variations in the depth of freezing, controlled by variations in snow-cover and sediment-water content, provide a possible explanation for the observed variation in tectonic style along the push-moraine. This suggestion is supported by the fact that the surge is known to have occurred in the late winter/early spring of 1999 and that some parts of the push moraine remain frozen today, while adjacent areas are unfrozen.

Similar variations in tectonic style could be caused by variations in the hydrogeology and therefore porewater pressures of the deforming sediment. However, although possible, variation in porewater pressures along the moraine are unlikely to provide a satisfactory explanation since the gross hydrogeological setting is similar along the length of the ridge. Some variation in the permeability of the pre-1998 sediment does occur, but there is no correlation between the outcrop of silty clay on the décollement surface and the location of areas of ductile deformation. Further, the sedimentology of the moraine is consistent along its length and that the total amount of strain undergone by the sediment along the 200 m length of the moraine is likely to have been consistent.

Cryospheric coupling during a winter surge

The landform-sediment assemblage associated with the 1998/1999 surge of Hagafellsjökull-Eystri fits well with the traditional surge-type landsystem (Evans & Rea 1999). However, there are a number of features that appear to reflect the seasonal timing of the surge. The first and perhaps most significant of these is the presence of evidence for subglacial deformation at sub-freezing temperatures. The association of subglacial sediment deformation with high pore water and low effective pressures within the substrate has led to the assumption that the process is restricted to areas where the glacier, or ice-sheet, bed is close to the pressure melting point and characterized by basal melting (Hart *et al.* 1990; Paterson 1994; Benn & Evans 1996; Boulton, 1996). The observations from Hagafellsjökull-Eystri add to the increasing weight of evidence to suggest that this is not always the case (e.g., Echelmeyer & Zhongxiang 1987; Astakhov *et al.* 1996; Cuffey *et al.* 1999, 2000; Fitzsimons *et al.* 2000). More importantly the data suggest that subglacial deformation may occur at sub-freezing temperatures, not just at polythermal and cold glacier in high altitude and latitude locations, but also around the margins of temperate glaciers during winter advances.

The second feature of interest is the detailed tectonic facies of the push moraine found along the eastern lateral margin of Hagafellsjökull-Eystri. Although the gross architecture of the moraine is consistent along its length, in detail the tectonic facies varies between sections that show brittle deformation and sections characterized by ductile deformation. The authors have argued that this may reflect spatial variation in the extent to which the glacier foreland was frozen prior to the ice advance. This is likely to be a feature that results from the timing of the surge. The observations also support those of Krüger (1993, 1994, 1996) and Matthews et al. (1995), who have already demonstrated the importance of seasonal freezing of glacier forelands to moraine formation.

The data from Hagafellsjökull-Eystri suggest that glacier surges that occur in winter/spring into seasonally frozen forelands may be associated with distinct modifications of the traditional surge-type landsystem. More importantly, the data presented here suggest that glaciers that advance into seasonally frozen forelands provide an opportunity to explore the coupling between glacier and permafrost.

The financial support of the Royal Society, NERC Geophysical Equipment Pool (loan 719), and the NERC Airborne Campaign (Iceland 2001) is acknowledged. Oddur Sigurðsson provided unpublished information and photographs of 1998/1999 surge. This work is covered by Rannis Research Declaration no. 5/2002. We would like to acknowledge the field assistance of Silvia Gonzalez, Alex Tomio, Simon Cook and Nick Midgely. Paul Zukowskyj prepared the digital elevation model in Fig. 3F, at the University of Hertfordshire, using data from the NERC Airborne Campaign (Iceland 2001).

References

ASTAKHOV, V. I., KAPLYANSKAYA, F. A., & TARNOGRADSKIY, V. D. 1996. Pleistocene permafrost of West Siberia as a deformable glacier bed. Permafrost and Periglacial Processes, 7, 165–191.
BENN, D. I. & EVANS D. J. A. 1996. The interpretation and classification of subglacially-deformed materials. Quaternary Science Reviews, 15, 23–52.
BENNETT, M. R., HUDDART, D. & MCCORMICK, T. 2000. An integrated approach to the study of glaciolacustrine landforms and sediments: a case study from Hagavatn, Iceland. Quaternary Science Reviews, 19, 633–665.
BENNETT, M. R., WALLER, R. I., MIDGLEY, N. G., HUDDART, D., GONZALEZ, S., COOK, S. J. & TOMIO, A. 2003. Subglacial deformation at sub-freezing temperatures? Evidence from Hagafellsjökull-Eystri, Iceland. Quaternary Science Reviews, 22, 915–923.
BOULTON, G. S. 1996. Theory of glacial erosion, transport and deposition as a consequence of subglacial sediment deformation. Journal of Glaciology, 42, 43–62.
BOULTON, G. S., VAN DER MEER, J. J. M., HART, J., BEETS, D., RUEGG, G. H. J., VAN DER WATEREN, F. M. & JARVIS, J. 1996. Till and moraine emplacement in a deforming bed surge – an example from a marine environment. Quaternary Science Reviews, 15, 961–987.
BROSTER, B. E. & CLAGUE, J. J. 1987. Advance and retreat glacigenic deformation at Williams Lake, British Columbia. Canadian Journal of Earth Science, 24, 1421–1430.
CUFFEY, K. M., CONWAY, H., HALLET, B., GADES, A. M. & RAYMOND, C. F. 1999. Interfacial water in polar glaciers, and glacier sliding at −17°C. Geophysical Research Letters, 26, 751–754.
CUFFEY, K. M., CONWAY, H., GADES, A. M., HALLET, B., LORRAIN, R., SEVERINGHAUS, J. P., STEIG, E. J., VAUGHN, B. & WHITE, J. W. C. 2000. Entrainment at cold glacier beds. Geology, 28, 351–354.
ECHELMEYER, K. & ZHONGXIANG, W. 1987. Direct observation of basal sliding and deformation of basal drift at subfreezing temperatures. Journal of Glaciology, 33, 83–98.
EVANS, D. J. A. & REA, B. R. 1999. Geomorphology and sedimentology of surging glaciers: a landsystem approach. Annals of Glaciology, 28, 75–82.
FITZSIMONS, S. J., LORRAIN, R. D. & VANDERGOES, M. J. 2000. Behaviour of subglacial sediment and basal ice in a cold glacier. In: MALTMAN, A. J., HUBBARD, B. & HAMBREY, M. J. (eds) Deformation of Glacial Materials. Geological Society Special Publication no. 176, London, 181–190.
HAMBREY, M. J. & HUDDART, D. 1995. Englacial and proglacial glaciotectonic processes at the snout of a thermally complex glacier in Svalbard. Journal of Quaternary Science, 10, 313–326.
HART, J. K., HINDMARSH, R. C. A. & BOULTON, G. S. 1990. Styles of subglacial glaciotectonic deformation within the context of the Anglian ice sheet. Earth Surface Processes and Landforms, 15, 227–241.
KRÜGER, J. 1993. Moraine-ridge formation along a stationary ice front in Iceland. Boreas, 22, 101–109.
KRÜGER, J. 1994. Glacial processes, sediments, landforms, and stratigraphy in the terminus region of Myrdalsjökull, Iceland. Folia Geographica Danica Tom., XXI, 233pp.
KRÜGER, J. 1996. Moraine ridges formed from subglacial frozen-on sediment slabs and their differentiation from push moraines. Holocene, 5, 420–427.
MATTHEWS, J. A., MCCARROLL, D. & SHAKESBY, R. A. 1995. Contemporary terminal-moraine ridge formation at a temperate glacier: Styggedalsbreen, Jotunheimen, southern Norway. Boreas, 24, 129–139.

M. R. BENNETT *ET AL.*

PATERSON, W. S. B. 1994. *The Physics of Glaciers*, 3rd edn, Pergamon Press, Oxford.

SHARP, M. J. 1985a. 'Crevasse-fill' ridges – a landform type characteristic of surging glaciers? *Geografiska Annaler*, **67A**, 213–220.

SHARP, M. J. 1985b. Sedimentation and stratigraphy at Eyjabakkajökull – an icelandic surging glacier. *Quaternary Research*, **24**, 268–284.

SHARP, M. J. 1988. Surging glaciers; geomorphic effects. *Progress in Physical Geography*, **12**, 349–370.

SIGBJARNARSON, G. 1975. Hagafelljökull eystri hlaupinn. *Jökull*, **26**, 94–96.

SIGURÐSSON, O. 2001. Jöklabreytingar 1930–1960, 1960–1990 og 1998–1999. *Jökull*, **50**, 129–136.

SOHEIM, A. 1991. The depositional environment of surging polar tidewater glaciers: a case study of the morphology, sedimentation and sediment properties in a surge affected marine basin outside Nordaustlandet, the northern Barents Sea. *Norsk Polarinstitutt Skrifter*, **194**, 97pp.

THEÓDÓRSSON, T. 1980. Hagafellsjöklar taka a rás. *Jökull*, **30**, 75–77.

WILLIAMS, P. J. & SMITH, M. W. 1989. *The Frozen Earth: Fundamentals of Geocryology*, Cambridge University Press, Cambridge.

Glacier–permafrost hydrological interconnectivity: Stagnation Glacier, Bylot Island, Canada

BRIAN J. MOORMAN

Department of Geography, University of Calgary, 2500 University Drive,
Calgary, Alberta, Canada T2N 1N4
(e-mail: moorman@ucalgary.ca)

Abstract: The complex thermal structure in areas where polythermal glaciers and continuous permafrost are present increases the potential for direct linkages between subsurface water conduits within glaciers and permafrost. In this study, hydrologic features of a glacier and the surrounding ice-cored moraines were examined and the potential for englacial water to flow out of the glacier and into the moraine was investigated. Ground-penetrating radar investigations, dye trace tests and direct observations of hydrological features (e.g., moulins, springs and caves) on and around Stagnation Glacier on Bylot Island, Arctic Canada, were undertaken. Data reveal that englacial conduits extend from the glacier into the adjacent ice-cored moraine. Glacial meltwater may have experienced variable flow conditions over the last 10 years and the conduit closures have occurred over a much longer time period. The study illustrates the interconnectivity of the glacial and permafrost hydrological systems.

In the high Arctic, where polythermal glaciers are surrounded by continuous permafrost, subsurface thermal conditions are frequently the dominant influences on the hydrological systems of both permafrost and glaciers. Permafrost conditions also enable glacial ice to be preserved (Moorman & Michel 2000a, 2003; Wolfe 1998) and for relict drainage structures to exist (Hodgkins 1997). As such, the hydrology of high Arctic glaciers and the surrounding permafrost may be more similar than lower latitude glaciers and their surrounding terrain. It is hypothesized that, in high Arctic periglacial ice-proximate environments, englacial hydrological conduits may be preserved in ice-cored moraines and the subsurface hydrological system can remain linked to the glacial system for extended periods of time. This hypothesis was tested on Bylot Island in the eastern Canadian Arctic, where caves (relict conduits) were discovered preserved in an ice-cored lateral moraine within permafrost terrain next to a polythermal glacier.

Englacial and subglacial hydrology is also fundamental to clarifying the influence of water in ice flow dynamics (Iken 1981; Willis 1995; Mair *et al.* 2002), as well as the development of glacier-induced floods, and the impacts of climate change on the glacier hydrological system. The evolution of glacial hydrological systems is central to how a glacier changes over time. However, the study of englacial and subglacial hydrology is hindered by the lack of accessibility. Techniques such as borehole imaging (Hubbard *et al.* 1995), dye tracing (Hooke *et al.* 1988), and ground-penetrating radar (Moorman & Michel 2000b) provide information at an instant in time, but the results are difficult to compare over a number of years.

Perennial reoccupation of moulins indicates that some aspects of the glacial subsurface hydrological system have a degree of stability over a multiyear time scale. Sedimentary structures, such as eskers and rounded debris on the end moraine, also provide evidence of englacial and subglacial hydrological activity (Kirkbride & Spedding 1996). However, the hydrological information that can be extracted from esker deposits is limited by post-depositional glacial and subaerial alteration. Significant advances in modelling subglacial water flow have recently been made (Clarke 1996; Richards *et al.* 1996); however, there are still many questions regarding the evolution of englacial systems (Benn & Evans 1998; Fountain & Walder 1998).

In high Arctic glacierized basins, the subsurface thermal conditions impact both the hydrological systems (Hodson & Ferguson 1999; Irvine-Fynn *et al.* 2004) and the preservation of buried ice following retreat

From: HARRIS, C. & MURTON, J. B. (eds) 2005. *Cryospheric Systems: Glaciers and Permafrost.*
Geological Society, London, Special Publications, **242**, 63–74.
0305-8719/05/$15.00 © The Geological Society of London 2005.

(Moorman & Michel 2000a). For example, some of the buried glacier ice on Bylot Island is estimated to be tens of thousands of years old (Klassen 1993) and, elsewhere, Paterson (1994) has shown that empty cavities in ice have the potential to be preserved over long periods of time due to the small creep rates associated with cold subsurface temperatures and low overburden pressures. Lunardini (1988), using radial heat flow models, has demonstrated that large water-filled cavities have the potential to be preserved for a number of years before freezing back due to the large heat capacity of water. This paper describes the character of caves discovered within an ice-cored lateral moraine, demonstrates how they were connected to the englacial hydrological system, and indicates they have the potential to be preserved and then reactivated in the permafrost environment.

Study area

Stagnation Glacier (B28 in Inland Waters Branch 1969) and its east lateral moraine were the focus of this study. The glacier is located on the southern portion of Bylot Island (Fig. 1). Bylot Island consists of a central ice field from which valley glaciers flow towards the coast. Glaciers on the island are currently at, or retreating from, their Neoglacial maximum positions that were attained within the last 100 years (Klassen 1993). Basal ice studies and ground-penetrating radar surveys indicate that at least some of the glaciers on the island, including Stagnation, are polythermal (Zdanowicz et al. 1996; Moorman & Michel 2000b).

Bylot Island falls well within the zone of continuous permafrost. The mean annual air temperature in the region is approximately $-15°C$. The average annual precipitation is less than 200 mm, with the maximum snow pack thickness being less than 80 cm in winter. From 3 years of shallow ground temperature measurements, the permafrost is estimated to be in the range of 200–400 m thick (Moorman & Michel 2000a). The active layer ranges from 30 to 50 cm in thickness. Retrogressive thaw slumps and thermokarst lakes on the island demonstrate that massive ice is present in the subsurface of recently exposed moraines and older sediments. Groundwater and subglacial springs have been identified in a number of locations across the island and large icings have been observed in front of at least six glaciers.

Stagnation Glacier is 9.0 km long and covers a total area of 14.1 km². Evidence from aerial photographs dating back to 1948 and

Fig. 1. A 1982 vertical aerial photograph of Stagnation Glacier drainage basin on southern Bylot Island, Canada. The location is shown by the star on the inset map. The white arrows indicate the location of caves in the east lateral moraine. Moulins are indicated by black circles, and springs are indicated by black dots

direct measurements since 1993 indicate that the glacier has retreated over 1.2 km since 1948 and is currently averaging a retreat rate of 4.9 m/a. The glacier is flanked by ice-cored moraines up to 70 m in height. The crest of the east lateral moraine varies in elevation from 380 m above sea level (a.s.l.) near the glacier terminus to 707 m a.s.l., 3 km up the valley where the moraine intersects the glacier surface.

Methods

A combination of ground-penetrating radar (GPR), dye tests and detailed cave mapping was used to investigate the hydrological character of the study site. In the summers of 1993–1996, 1999, 2001 and 2002, hydrological features of Stagnation Glacier were studied and mass movements on the lateral moraines exposed abandoned, and sometimes still active, subsurface hydrological conduits. These relict conduits (and their external manifestations as caves) were mapped and their relationship to elements of the glacier hydrological system (e.g., moulins, englacial drainage networks, springs and emerging eskers) was determined.

Ground-penetrating radar

Ground-penetrating radar was used to map the thickness and internal structure within the glacier and the surrounding moraines. Extensive surveys were conducted on the glacier by fixing the antennas together and dragging the GPR system over the ice. Repeated surveys were conducted over the lower 3 km of the glacier in 1993, 1994, 1999 and 2002. The greatest density of surveys was undertaken in 2002, where a grid with three GPR lines parallel to the glacier long axis and GPR cross-lines every 50–100 m up from the terminus of the glacier was repeated three times throughout the summer season. The base of the glacier and the location of englacial voids and thermal interfaces were mapped with the GPR. The polarity of the GPR reflections was used to determine whether the voids were water- or air-filled. Surveying the moraines was considerably more difficult as the antennas had to be moved over the debris cover in a step-wise fashion 1 m at a time. Six GPR surveys from the glacier's edge up and over the lateral moraines were conducted at selected locations. The coarse debris on the surface of the moraines produced a considerable amount of noise on the GPR profiles; however, the subsurface structures were still discernable. The details of the GPR profiling and interpretation are discussed in Moorman & Michel (2000b). Because of the excellent propagation properties of ice and frozen sediment, the GPR was routinely able to image to depths over 100 m.

Dye tracing

Dye tracing was used to determine the connectivity between where water entered the subsurface hydrological system and where it re-emerged. Dye slugs were introduced into streams on the glacier surface and surrounding permafrost terrain as they proceeded underground. As the majority of the travel path of the dye was in the subsurface, photodegradable flouroscein dye was used. Initial tests were conducted in the lateral stream to determine the appropriate amounts of dye to use and maximum velocities to expect. The velocity of the surface was up to 19 m/s, while subsurface water flow ranged from 0.16 to 0.32 m/s. It was determined that 500 g dye slugs would ensure visual identification of the plume where there was no significant long-term subsurface storage.

In most cases dye recovery was very obvious and quantitative analysis showed a greater than 90% dye recovery in most tests. In a few cases, injected dye was not observed to return to the surface even at part-per-million levels after many days. To rule out a high dilution factor resulting in not being able to detect the discharge of dye, these tests were repeated with dye slug quantities up to five times larger.

Cave surveys

The location and elevation of four caves discovered in the east lateral moraine of Stagnation Glacier were mapped (Figs 1 and 2). Their size and internal structure were repeatedly measured. However, detailed internal surveys were not always possible due to the unstable nature of the glacial debris cover on the moraine. Seven caves in the east side of the glacier were also studied; however, due to their transitory nature, none of the glacier caves were observable for more than one year.

Results

Extensive GPR surveys and repeated summer observations of the study site have resulted in a comprehensive data set on the structural relationship between Stagnation Glacier and the massive ice within the surrounding permafrost, englacial hydrology and the character of caves in the margins of the glacier and surrounding moraines.

Glacier and moraine structure

Ground-penetrating radar surveys across the glacier and lateral moraines show that the moraines are ice-cored and that the ice core is connected to the glacier ice mass (Fig. 3). Located roughly 300 m from the 1993 terminus of the glacier, the GPR profile shows the glacier to be up to 71 m thick in the centre and 21 m thick at

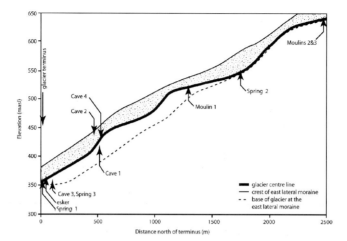

Fig. 2. The long profile of Stagnation Glacier showing the location of hydrological features relative to the crest of the east lateral moraine, the crest of the glacier (roughly at its centre line) and the base of the glacier where it meets the east lateral moraine

its eastern margin. From the GPR data, and observations at the glacier margins, there appears to be a 1–5 m basal ice layer. The basal ice observed in the end moraine and cave 1 consisted of dispersed, fine-grained sediment with occasional cobbles scattered throughout.

The ice core within the lateral moraine was on average 15 m thick, pinching out at the top of the moraine. The ringing on the GPR profile in Fig. 3 at the glacier margin is caused by the lateral

stream and is not representative of subsurface features. The average thickness of the glacial debris covering the ice core of the moraine was measured as 0.75 m, being slightly thicker near the base of the moraine as a result of mass movement.

A 4 m thick ice core within the end moraine extending up to 200 m beyond the end of the glacier was also mapped with GPR. This proglacial buried ice is contiguous with the ice

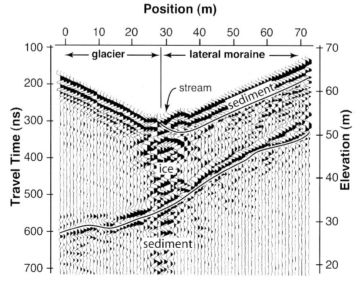

Fig. 3. A GPR profile across the east side of Stagnation Glacier and up the east lateral moraine displays the connection between the glacier body and the ice core of the moraine

core of the lateral moraines, resulting in a continuous fringe of buried ice around the glacier (Moorman & Michel, 2000b).

Englacial hydrology

There are several lines of evidence that indicate englacial water flow within Stagnation Glacier. Englacial drainage networks were mapped from GPR surveys undertaken in 1993, 1994, 1999 and 2002. Verification of this GPR interpretation was accomplished at a large icing in front of the adjacent glacier where conduits within the icing could be directly observed at the icing edge or could be cored into (Moorman & Michel 2000b).

Direct evidence of englacial flow is provided by the presence of moulins and emerging eskers at the glaciers margin. Moulin 1 has been stable and experiencing repeated reactivation each summer since at least 1999 and two other moulins have been observed within the ablation area of the glacier during hot summers (Fig. 1). During colder periods, there is not sufficient supraglacial runoff to activate all of the moulins. The moulins have also been observed to become temporarily plugged by slush flows after summer snowfalls. The combination of the slush plug and mounding of slush around the entrance to the moulin diverts supraglacial water flow away from the entrance. These slush plugs can last for days to weeks, depending on the weather.

An ice-cored esker has been observed on aerial photographs to be emerging from the terminus of the glacier over the last 20 years (Fig. 2), indicating the englacial flow of sediment-laden water in the past. In the proglacial region there was up to 16 m of ice beneath 2 m of well sorted esker sand. The esker sediment stringer was also mapped with GPR to extend over 100 m back into the glacier at an elevation 10 m above the bed. The presence of clean ice beneath the esker sediment indicates that the conduit that formed the esker was above the basal zone of the glacier.

Flowing springs were observed on the surface of the glacier, or at its margins each summer the study area was visited. An icing forms at the terminus of the glacier each winter with icing blisters developing close to the glacier margin each year. In 2002, the icing covered roughly 30,000 m^2 and was up to 2.5 m thick. Three icing blisters formed within 5 m of the glacier margin. From June 10, until June 24 2002 all of the icing blisters were observed to be covered by an intact ice surface. On June 25, clear water started emerging from spring 1 on the largest blister and continued for the next week, with a flow estimated to be 0.35 m^3/s (S1 in

Figs 1 and 4). The water was clear with no appreciable suspended sediment (Table 1), indicating that the englacial water feeding this blister never reached the sediment-rich base of the glacier.

Over a 24 h period between July 2 and 4 2002, sediment-laden water emerged at the surface of the glacier from spring 2 approximately 1.8 km from the glacier terminus (S2 in Figs 1 and 5). The spring emerged from a number of locations along a fracture in the ice with the majority of the water coming out of two main conduits located with 25 m of each other. The suspended sediment concentration of the spring water was appreciably greater than that of the proglacial stream at this time (Table 1) or any other water sources. This suggests that the source of the water had to have been the basal ice layer or the sediment at the base of the glacier.

Caves

Four caves have been mapped in the east lateral moraine near the terminus of Stagnation Glacier. These relict conduits extended deep into the ice core of the moraine; some were over 2 m in diameter for up to 55 m when they extended along the long axis the moraine. The walls of the caves consisted of basal ice or glacial ice, or frozen glaciofluvial deposits. Cave 1 (C1 in Fig. 1) formed on the outside of the moraine and is visible on aerial photographs extending back to 1948. Over that period, the cave grew to be over 100 m wide; in 2001 the roof of cave 1 collapsed and all that remains today is a notch cut through the lateral moraine. Before 2001, cave 1 had a large cavernous entrance which rapidly decreased in size to less than 1 m in diameter within 30 m of the entrance.

Over the last 10 years, the terrestrial stream that runs along the outside of the lateral moraine has drained into this cave, towards the glacier. The source of this stream is a 3.75 km^2-high relief drainage basin. The basin contains two sub basins; one is ice free and the other contains a small cirque glacier. During the summer melt season the stream entering this cave has an average discharge in the range 0.2–1.0 m^3/s. However, large rain storms in 2002 demonstrated that during peak flow the stream discharge can increase by an order of magnitude for a day or two. The onset and cessation of these peak flows occurred so quickly that the conduit in cave 1 could not enlarge fast enough to carry all of the water. Dye tracing tests in 1994 indicated that the stream entering the cave was not connected to the ice marginal stream, but

Fig. 4. The clear discharge from an icing blister 5 m from the edge of Stagnation Glacier (spring 1) in June 2002. The glacier terminus is 4 m to the left and the east lateral moraine can be seen in the background

Table 1. *Suspended sediment concentrations for water sources around Stagnation Glacier*

Water source	Suspended sediment concentration (g/l)
Proglacial stream 2002 average	0.25
Proglacial stream 2002 peak	3.25
Proglacial stream July 2–4 range	0.4–0.6
Spring 1	0.0
Spring 2, July 2–4	0.4
Supraglacial streams	0.0

the water was being stored in the subsurface. After considerable erosion and down cutting, when the cave collapsed in 2001, a complex network of small conduits (less than 1 m diameter) was exposed in the walls at three different levels. Dye trace test conducted in 2002 showed that the stream entering the cave emerged from within the glacier approximately 100 m down valley from the cave joining the east lateral stream.

Cave 2 (C2 in Fig. 1) was exposed from 1999 until 2001. It was 30 m down valley from cave 1 on the glacier side of the moraine. This cave

Fig. 5. The water from spring 2 on July 3 2002, emerged from two proximal main conduits. Each of the sources had identical suspended sediment concentrations that were appreciably higher than that of the proglacial stream at that time

became exposed when an active layer detachment eroded the side of the moraine, revealing the side of a conduit. This conduit ran parallel to the lateral moraine, then inclined down valley for approximately 15 m before heading east, deeper into the core of the moraine. This cave extended roughly 30 m into the moraine before it became impassible. The upper most portion of cave 2 was close to 7 m wide but its height was as little as 0.5 m. The first 10 m of the cave was gently inclined, the next 13 m was a steeply dipping set of frozen rapids with a boulder-lined floor which then opened up into a large chamber over 4 m in height and 8 m across. Basal ice was exposed in the walls of the cave, containing diffuse sediment inclusions ranging from clay to cobble sized particles.

Cave 3 (C3 in Fig. 1) was exposed only during the summer of 1999. However, for over 3 weeks in July 1994, spring 3 emerged from the debris covered moraine at the same location and was flowing at rate of approximately $2-4 \, m^3/s$ (Fig. 6). This cave was on the glacier side of the moraine near the base of the moraine at the glacier terminus. Cave 3 had a oval cross section and was still 2 m high 55 m in from the entrance as the cave curved up the valley, gradually increasing in elevation.

Cave 3 had a small trickle of water draining out of it throughout July 1999, with evidence that much greater flows had occurred recently.

Just outside the entrance to the cave 2–4 m diameter blocks of glacier ice were lying on the ground with random orientations. There were 1–2 m blocks of glacier ice on the floor of the cave 10 m from the entrance at a bend in the cave route. Their sub-rounded shapes and variable orientation indicated that they had also been deposited by water.

Decimeter-sized blocks of glacier ice were frozen to the ceiling of cave 3 (Fig. 7). These blocks were frozen in place by a 10–15 cm thick veneer of ice that had frozen to the cave walls. This layer of ice was continuous from floor to ceiling and extended deep into cave. Near the cave opening, sunlight differentially melted the ice to reveal its layering (Fig. 8). The freezing of water on the ceiling indicates that the cave was full of water, experiencing low velocity or stagnant conditions during a period with a negative energy balance (i.e., winter). The layering of the ice veneer suggests that there may have been several episodes where the cave was full of water that was freezing for a period of several weeks to months.

Cave 4 (C4 in Fig. 1) was exposed in 2001 as cave 1 expanded and eroded further into the ice core of the moraine. Its opening was located on the north side of cave 1, on the glacier side of the crest of the moraine, extending further into the heart of the moraine towards the

Fig. 6. The water flowing from spring 3 in July 1994, on the east lateral moraine at the location where cave 3 would later be exposed. Note person for scale

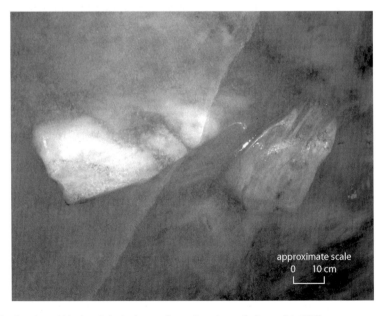

Fig. 7. Randomly oriented blocks of glacier ice are frozen into the roof of cave 3 in 1999

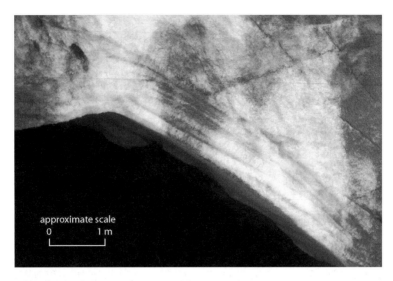

Fig. 8. At the entrance to cave 3 in July 1999 layers of ice that had frozen on to the walls of the cave while it was filled with water can be seen

glacier, gradually curving up the valley (Fig. 9). At its opening, the cave was 6 m high and 5 m wide, gradually decreasing to 3 m in diameter 50 m from the entrance. At this location a plug of ice prohibited further passage. Cross section maps of the cave indicate that it was 75% full of water when refreezing was initiated. The majority of the water closer to the entrance was able to drain into a lower level branch conduit before freezing, thus leaving the floor of the cave covered by a 15–25 cm deep layer of water that had a thin (1–2 cm) ice cover on it near the cave entrance. This horizontal floor of ice was completely frozen deeper into the cave. The only aggradational ice on the walls of this cave was hoar frost.

Fig. 9. Hoar frost coats the walls of cave 4 within the east lateral moraine, July 2001, while a thin layer of ice covers water remaining in the bottom of the cave

The englacial caves at the terminus and along the eastern margin on the glacier ranged from less than 1 m in diameter to over 5 m. They all had evidence of past water flow. In some caves there was still small some water flowing through them. Several of the caves in the side of the glacier had complex shapes, indicating multiple stages of phreatic flow. Caves were contained within buried glacier ice and basal ice.

Discussion

Connectivity between the glacier and lateral moraine

Repeated GPR surveys revealed the presence of englacial conduits that were water-filled in the summer. The GPR interpretation indicated that some englacial conduits carried water to the glacier's edge without reaching the basal zone. This was confirmed by the sediment-free water emerging from spring 1.

The hydrological system within the glacier also contained conduits that reached the bed before returning to the glacier surface, as demonstrated by sediment-rich water emerging from spring 2. The elevation of the moulins up the valley from spring 2 was in the order of 100 m higher than the glacier surface in the vicinity of spring 2. The glacier was 120 m thick in this area. Meltwater routing from moulins 2 and 3 down to the glacier bed and rising back to the surface along fractures would account for the location of the spring (considering the elevation differential) and sediment content of the emerging water. After the flow from this spring ceased, dye tracing tests on moulins 2 and 3 showed that they subsequently emerged at the glacier terminus. This attests to how rapidly the en/subglacial drainage can become rerouted.

The constant summer flow of the terrestrial stream from the basin beyond the east lateral moraine was clearly the dominant factor in the recent evolution of cave 1. This is displayed on the aerial photographs dating back to 1948. However, the sizes of caves 2 and 3 suggest that

much greater volumes of water flowed through them than is currently generated by the terrestrial stream flowing into cave 1. The most likely source of this water would not be the terrestrial stream or groundwater but the glacier itself. The elevation of cave 2 is well above the current elevation of the glacier and ice-marginal stream (Fig. 2); thus it is unlikely that stream, or englacial, water has occupied cave 2 recently. However in the past, when the glacier was larger and the ice core of this lateral moraine was still part of the glacier, flow from central portions of the glacier into cave 2 would be quite reasonable. The large width to height ratio of cave 2 also suggests that it has been abandoned, and been subjected to non-uniform ice creep closure for quite some time (Table 2). The roundness of the other three caves indicates that they have recently been subject to phreatic flow.

Extensive dye trace tests in 1994 indicated that spring 3 (i.e., cave 3) was not connected to the terrestrial stream entering cave 1. There was also no other terrestrial source large enough to feed spring 3. It is also not feasible that a large enough storage cavity existed within the moraine to keep the spring flowing throughout the summer, at the rate observed. Thus the source of the water feeding that spring is most likely the glacier. To be so well connected to the glacial hydrological system, this conduit was probably part of the glacier's englacial hydrological network when the glacier was considerably larger, and this ice was still part of the glacier. This conduit has probably been preserved since the last Neoglacial maximum, being reactivated in 1994.

The source of water that formed cave 4 is unclear. Its size and elevation suggest that it was connected to a larger source of water such as the englacial or subglacial hydrological system. It is also possible that it may have been a past route for the terrestrial stream to make its way through the moraine and into the glacier. However, it is much larger than would be created by the current average summer discharge of the terrestrial stream that enters cave 1. Given the

Table 2. *Characteristics of caves in the east lateral moraine of Stagnation Glacier*

Cave number	Cave elevation (m a.s.l.)	Moraine elevation (m a.s.l.)	Water flow direction	Years exposed	Cross section (m^2)	Width/height ratio
1	400	Eroded	In	<1948–present	0.8	1.8
2	425	440	In	1999–2001	7.0	14
3	330	420	Out	1999[a]	50.3	1.9
4	410	444	In	2001–present	23.8	0.7

[a]A spring emerged from the location of cave 3 as early as 1994.

characteristics of caves 2–4, it is apparent that past englacial activity has played a role in the development of conduits that are currently preserved within this lateral moraine.

Conduit preservation

As proposed by Hodgkins (1997), the characteristics of polythermal glaciers make it possible to preserve relict drainage systems that were created under past climatic conditions. The evidence presented here indicates that relict conduits are not only preserved within the glacier, but also in the ice-cored moraines left behind after glacial retreat. As well, the characteristics of these caves indicate that the flow of water from the glacier out into the moraine, and vice versa, not only occurred in the past but continues today.

The current size and retreat rate of the glacier indicate that the conduits in the lateral moraine have been preserved for over 50 years. Cold ground temperatures and the relatively small overburden pressure have resulted in very slow creep rates and good preservation of empty conduits. Once the ice is separated from the main body of the glacier and buried in the moraine, it is insulated by the sediment cover and will cool until it reaches thermal equilibrium with the local climate. Data reported in Paterson (1994) indicate that the flow parameter decreases by an order of magnitude between 0 and $-10°C$. The $-9.6°C$ mean annual ground temperature recorded in the area would greatly reduce the ability of the ice to creep. This would significantly slow the closure of empty conduits in moraines. If the conduit is filled with water (flowing or not), creep is slowed due to increased hydrostatic pressure.

The thermal structure of Stagnation Glacier varies seasonally as well as inter-annually. The GPR surveys indicate that the lateral extent of water-rich ice at the pressure melting point can change by over 200 m in 2 months. This is indicative of the complex and rapidly changing hydro-thermal structure of the glacier. As such, conduits that are initiated in thermal equilibrium with the surrounding ice may quickly find themselves in a state of disequilibrium. However, owing to the thermal momentum of flowing water, the routing may be preserved for a considerable period. Water flowing through a conduit in cold ice will create a bulb of warm ice around the conduit, retarding freeze-back.

Freeze-back of the conduit walls will occur when it contains stagnant water that can result in closure of the conduit, as was observed to a small degree in cave 3. However, the rate of freeze-back is limited by the large latent heat of fusion for water, and the thermal conductivity and thickness of the surrounding ice and sediment. Unlike surface water bodies, heat loss is only through conductive heat flow. Using the cylindrical heat flow concepts developed by Lunardini (1988) and a constant heat loss to the surrounding ice, it was conservatively estimated that a 1 m diameter conduit would take over a year to refreeze and a 5 m diameter conduit would take over 30 years in this setting. Therefore, short-term conduit preservation is favoured in this environment.

Over longer periods of time there is the possibility that the conduit will either be reactivated or drain. The subsurface hydrological system at this site is temporally dynamic and the drainage routing at any given time is strongly influenced by current meteorological and hydrological conditions. Thus reactivation of relict conduit may occur quickly and last for a short period of time, having little long-term impact on the conduit evolution.

Conclusions

In studying the hydrological features and imaging the subsurface of Stagnation Glacier with GPR, it has been shown that there is active englacial flow. It was also revealed that the ice core of the east lateral moraine is still connected to the main ice body of the glacier. Englacial conduits extend beyond the edge of the glacier into the buried ice in the surrounding moraines. A spring carrying subglacial water to the surface of the glacier indicates that high hydraulic heads are generated within the glacier.

The presence of caves in the lateral moraine are indicative of preserved relict englacial conduits from periods when the glacier was larger; they have been preserved in the lateral moraine and are occasionally reactivated. The thermal properties of water and the structural and thermal properties of the surrounding ice enable the conduits to be preserved for extended periods of time, depending on the nature of the water flow regime within the conduit. It is also suggested that the routing and preservation of the englacial hydrological system is largely a function of the complex thermal and hydrological setting in polythermal glaciers and the surrounding permafrost. The three dominant controls in conduit preservation within buried ice are their depth beneath the surface, the temperature of the permafrost, and the time since burial. The insulating effect of sediment covering buried ice causes buried ice close to the surface to cool, slowing the rate of creep. Thus the preservation of subsurface conduits is enhanced by the transition from glacial to permafrost terrain.

This research would not have been possible without the field and laboratory assistance of M. Elver, D. Kliza, L. Moorman, C. Deacock, A. Lyttle, J. Williams, F. Walter and T. Irvine-Fynn. The Polar Continental Shelf Project provided logistical assistance for this project. The Northern Science Training Program and the Natural Sciences and Engineering Council of Canada provided financial assistance for field work. The Hamlet of Pond Inlet is acknowledged for their support. Critical review of the manuscript by T. Irvine-Fynn, R. Koerner and an anonymous referee is also very much appreciated and led to the improvement of the text. Finally, the advice, knowledge, friendship and field assistance of F. Michel is very greatly appreciated.

References

BENN, D. I. & EVANS, D. J. A. 1998. *Glaciers and Glaciation*, Arnold, London.

CLARKE, G. K. C. 1996. Lumped-element analysis of subglacial hydraulic circuits. *Journal of Geophysical Research*, **101**(B8), 17547–17559.

FOUNTAIN, A. G. & WALDER, J. S. 1998. Water flow through temperate glaciers. *Reviews of Geophysics*, **36**, 299–328.

HODGKINS, R. 1997. Glacier hydrology in Svalbard, Norwegian High Arctic. *Quaternary Science Reviews*, **16**, 957–973.

HODSON, A. J. & FERGUSON, R. I. 1999. Fluvial suspended sediment transport from cold and warm-based glaciers in Svalbard. *Earth Surface Processes and Landforms*, **24**, 957–974.

HOOKE, R. L. B., MILLER, S. B. & KOHLER, J. 1988. Character of the englacial and subglacial drainage system in the upper part of the ablation area of Storglaciären, Sweden. *Journal of Glaciology*, **34**, 228–231.

HUBBARD, B. P., SHARP, M. J., WILLIS, I. C., NIELSEN, M. K. & SMART, C. C. 1995. Borehole water-level variations and the structure of the subglacial hydrological system of Haut Glacier d'Arolla, Valais, Switzerland. *Journal of Glaciology*, **41**, 572–583.

IKEN, A. 1981. The effect of subglacial water pressure on the sliding velocity of a glacier in an idealized numerical model. *Journal of Glaciology*, **27**, 407–421.

INLAND WATERS BRANCH 1969. *Glacier Atlas of Canada: Bylot Island Glacier Inventory Area 46201*, Department of Energy, Mines and Resources, Ottawa.

IRVINE-FYNN, T. D. L., MOORMAN, B. J., WILLIS, I. C., SJOGREN, D. B., HODSON, A. J., MUMFORD, P. N., WALTER, F. S. A. & WILLIAMS, J. L. M. 2004. Geocryological processes linked to high-arctic pro-glacial stream suspended sediment dynamics: Examples from Bylot Island, Nunavut and Spitsbergen, Svalbard. *Hydrological Processes* (in press).

KIRKBRIDE, M. & SPEDDING, N. 1996. The influence of englacial drainage on sediment-transport pathways and till texture of temperate valley glaciers. *Annals of Glaciology*, **22**, 160–166.

KLASSEN, R. A. 1993. Quaternary geology and glacial history of Bylot Island, Northwest Territories. Memoir, 93 pp.

LUNARDINI, V. J. 1988. Heat conduction with freezing or thawing. US Army Cold Regions Research and Engineering Laboratory, Hanover, NH, 329 pp.

MAIR, D., NIENOW, P., SHARP, M. J., WOHLLEBEN, T. & WILLIS, I. 2002. Influence of subglacial drainage system evolution on glacier surface motion: Haut Glacier d'Arolla, Switzerland. *Journal of Geophysical Research*, **107** (B8), 10.1029/2001JB000514.

MOORMAN, B. J. & MICHEL, F. A. 2000a. The burial of ice in the proglacial environment on Bylot Island, Arctic Canada. *Permafrost and Periglacial Processes*, **11**, 161–175.

MOORMAN, B. J. & MICHEL, F. A. 2000b. Glacial hydrological system characterization using ground-penetrating radar. *Hydrological Processes*, **14**, 2645–2667.

MOORMAN, B. J. & MICHEL, F. A. 2003. Burial of glacier ice by deltaic deposition. *In*: PHILLIPS, M. SPRINGMAN, S. M. & ARENSON, L. (eds) *Permafrost: 8th International Conference*, Zurich, 21–25 July 2003, Balkema, Rotterdam, 777–782.

PATERSON, W. S. B. 1994. *The Physics of Glaciers*, Pergamon Press, New York.

RICHARDS, K., SHARP, M., ARNOLD, N., GURNELL, A., CLARK, M., NIENOW, P., BROWN, G., WILLIS, I. & LAWSON, W. 1996. An integrated approach to modelling hydrology and water quality in glacierized catchments. *Hydrological Processes*, **10**, 479–508.

WILLIS, I. C. 1995. Intra-annual variations in glacier motion: a review. *Progress in Physical Geography*, **19**, 61–106.

WOLFE, S. A. 1998. Massive ice associated with glaciolacustrine delta sediments, Slave Geologic Province, N.W.T. Canada. *In*: LEWKOWICZ, G. & ALLARD, M. (eds) *Permafrost: Seventh International Conference*, Collection Nordicana no. 57, Centre d'études nordiques, Université Laval, Yellowknife, 1133–1139.

ZDANOWICZ, C. M., MICHEL, F. A. & SHILTS, W. W. 1996. Basal debris entrainment and transport in glaciers of southwestern Bylot Island, Canadian Arctic. *Annals of Glaciology*, **22**, 107–113.

Glacier–rock glacier relationships as climatic indicators during the late Quaternary in the Cordillera Ampato, Western Cordillera of southern Peru

UWE DORNBUSCH

*Department of Geography, Chichester Buildings, University of Sussex, Falmer,
Brighton BN1 9QJ, UK
(e-mail: u.dornbusch@sussex.ac.uk)*

Abstract: Mapping of glacial and periglacial features along a 15 km long, north–south trending ridge at the eastern end of the Cordillera Ampato (Fig. 1) has been carried out using aerial photograph interpretation. Over the length of the ridge a distinct change in features from moraines and small rock glaciers in the north to large rock glaciers and a lack of moraines in the south can be observed. It is suggested that this feature change reflects a steep precipitation gradient during the Pleistocene. Comparison with geomorphological mapping – both in the field and from air photographs – in other areas of the Cordillera Ampato and further west shows that this feature change is unique and that its location at the eastern end of the Cordillera is in good agreement with present-day precipitation distribution. This indicates that the general precipitation pattern, and thus the general circulation pattern, during the Pleistocene was very similar to that today.

The Western Cordillera of southern Peru has long been neglected as an area for Quaternary research, despite its ideal location to study climate change. Apart from the political situation up until the early 1990s the remoteness of many areas and the lack of access has made field work very difficult. Over 5500 m above sea level (a.s.l.), recently or presently glaciated peaks of the Cordillera Ampato (Fig. 1) can be found in close proximity to the coastal desert that reaches up to 2000 m a.s.l. along its western flanks (Abele 1992). Modern general atmospheric circulation transports moisture from the Amazon basin into the Andes, facilitated by deeply incised valley systems that reach deep into the mountains from the north-west. This leads to a reduction in precipitation from northwest to southeast (Hoffmann 1975; Dornbusch 1998), which is reflected both in terms of changes in the snowline altitude (e.g., Nogami 1982; Fox 1993; Dornbusch 1998), and a steep precipitation gradient along the western edge of the Western Cordillera that is responsible for abrupt vegetation changes (Weberbauer 1922). If these gradients can be found in the same regions in the Quaternary record, they are likely to indicate similar general circulation patterns and may be used to evaluate claims of latitudinal shifts of the circulation belts during the Pleistocene (e.g., Heusser 1989; Markgraf 1993; Messerli et al. 1993).

Detailed map and aerial photograph analysis, together with field mapping, provide detailed information on Holocene and Pleistocene glacier extents and active and fossil rock glaciers along the western margin of the Western Cordillera in southern Peru (Dornbusch 1997, 2000, 2002) where ages of the glaciations and cold phases thought to have been responsible for their formation have been assigned purely on positional relationship and correlation with similar features in other parts of the Andes (Dornbusch 2000; Table 2).

Local climatic data have been compiled by Dornbusch (1998) and combined with present-day snowline data to assess the present-day climate–snowline altitude relationship.

Based on the present relation between climate (precipitation and height of the 0°C isotherm) and snowline altitude and that of climatic conditions inferred from Pleistocene snowlines and accumulation area ratios, Dornbusch (1997, 2002) concluded that the western edge of the Western Cordillera in southern Peru was influenced by similar, if not moister, conditions compared with today and that no indications of an increased aridity could be found, contrary to the findings of Fox (1993) or Klein et al. (1999).

While active and (most) fossil rock glaciers follow the snowline trend with a rise from west to east in the Cordillera Ampato (Dornbusch 2002), a peculiar assemblage of Pleistocene

From: HARRIS, C. & MURTON, J. B. (eds) 2005. *Cryospheric Systems: Glaciers and Permafrost.*
Geological Society, London, Special Publications, **242**, 75–82.
0305-8719/05/$15.00 © The Geological Society of London 2005.

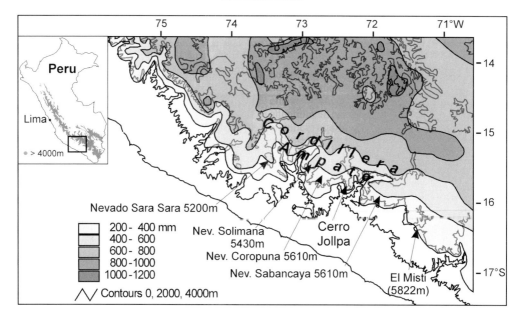

Fig. 1. Mean annual precipitation map for part of the Western Cordillera of southern Peru (after Dornbusch 1998). Names refer to glaciated summits with the height of the mean snowline altitude given, except for the unglaciated El Misti where the summit altitude is given

moraines and fossil rock glaciers can be found in the eastern part of the range along the Cerro Jollpa ridge, indicating a very steep moisture gradient that has produced exceptionally large rock glaciers.

This paper will briefly analyse the present climate–glacier/rock glacier relationship in the Cordillera Ampato (Dornbusch 1998), summarize the findings of Dornbusch (2002) on the active and inactive rock glaciers in the region as derived from aerial photographs, describe the unique feature change from moraines and small rock glaciers in the north to large rock glaciers and a lack of moraines in the south along the Cerro Jollpa ridge and discuss the palaeoclimatic implications.

Definitions

The author has an open mind on the origin of the features commonly named rock glaciers (Whalley & Azizi 2003) and is of the opinion that debris-covered rock-ice flows (Clark *et al.* 1994) can originate purely under permafrost conditions or as the result of a glacier being buried under surface debris, and that a transition between the ice-cored and the ice-cemented rock glacier is possible (Clark *et al.* 1994) both on a spatial (e.g., the 'rooting zone' model of Barsch & King 1989; Domaradzki 1951) and temporal scale. For each rock glacier, one main form of

origin seems to be more likely than another; however, the author is aware that the interpretations given in this paper may be controversial.

The common subdivision of rock glaciers into active, inactive and fossil based on morphological criteria (Wahrhaftig & Cox 1959; Martin & Whalley 1987) is applied to rock glaciers in this study.

Present-day climate–glacier/rock glacier relationship

Mean annual precipitation values for a large number of observation stations together with isohyets from a variety of sources were compiled into a precipitation map of south Peru by Dornbusch (1998). Although most of the stations on which the map is based were only operational from the late 1960s, comparison with stations with longer time series and long-term precipitation proxy data (river discharge) shows that the isohyets slightly underestimate the average annual precipitation for the period covering the 1930s to the 1970s.

Figure 1 is a detailed view of the precipitation map focusing on the Cordillera Ampato (Dornbusch 1998). The main precipitation trend is a decrease from north to south with precipitation values north of the continental watershed generally exceeding 800 mm. While the valley heads reaching up to the watershed from the north bring in moist air from the Amazon basin, those reaching up from the Pacific transport drier air further inland. As a consequence, the isohyets on the Pacific side follow the general topography, reaching further inland where large valleys have cut deep into the Western Cordillera. However, they do not follow the contours, but run obliquely to them; for example, the 200 mm isohyet rises from as low as 2000 m a.s.l. in the northwest to 4000 m a.s.l. in the southeast.

Based on the Peruvian Glacier Inventory (Ames *et al.* 1988), Dornbusch (1997, 1998) calculated the regional and local mean snowline altitudes for southern Peru. The height of the regional snowlines for the Nevados Sara Sara, Solimana, Coropuna and Sabancaya are shown in Fig. 1. The rise in snowlines coincides with a decrease in precipitation, while the height of the 0°C isotherm remains at the same level of *c.* 4900 m a.s.l. in the same region (Dornbusch 1998), indicating that precipitation, rather than temperature, is the main variable controlling the height of the snowline, as it is further south in Chile (Messerli *et al.* 1993; Grosjean *et al.* 1996). This can be further illustrated at El Misti (Fig. 1), which is not glacierized at present (Silva & Francis 1991) and has not been in the recent past (Hastenrath 1967). The small amount of annual precipitation at this locality has raised the snowline above the summit.

Few presently active rock glaciers exist in the Cordillera Ampato. Field observations by Dornbusch (1997, 2000) at Nevado Sara Sara have documented active rock glaciers along its western and southern side, while aerial photograph interpretation has found one on the northern flank. All show typical signs of active rock glaciers with a surface morphology characterized by ridges and v-shaped furrows, coarser blocks on top and unvegetated slopes. Slope angles of the snouts are at the angle of repose (between 33 and 35°) and facilitate rock falls observed in the field. These rock glaciers are generally small, rarely exceeding *c.* 200 m in horizontal length, but can have quite high (up to 80 m) snouts. All of the rock glaciers are situated above the maximum elevation of any moraines, with the tops of the snouts on Nevado Sara Sara lying between 4550 and 4800 m a.s.l. on the western and southern slope and at *c.* 5100 m a.s.l. on the northern slope. Some of these rock glaciers merge into glaciers higher up while others have no visible glacier above them but are linked by debris cones and aprons to the rocky back wall. Their size, and the lack of glacial moraines on those which are still backed by a glacier, seem to indicate a relatively young age (probably 'Little Ice Age') and a glacial origin (i.e., they are more likely to be ice-cored rock glaciers).

Below the active rock glaciers, a smaller number of fossil rock glaciers can be found with snouts reaching down to *c.* 4500 ma.s.l., which are comparable in size to the active rock glaciers; all of them lie in areas that show glacial activity to well below that altitude.

It is assumed that rock glaciers on Nevado Sara Sara formed not as a consequence of increasing aridity, but in response to shrinking glaciers caused by a rise in the ELA that exposed larger areas of the bedrock which contributed to a supra-glacial sediment cover, similar to the model proposed for the Reichenkar Rock glacier (Krainer & Mostler 2000).

Aerial photograph interpretation of Nevado Solimana indicates nine active rock glaciers that are up to 1 km long with snout altitudes from 5100 m a.s.l. on the southern flank to 5450 m a.s.l. on the northern flank (Dornbusch 2002). Below these, over 30 fossil rock glaciers can be found which are much smaller and generally located in the shadow of a minor back wall. The lowest have snout altitudes of 4500 m a.s.l. in southern but can reach up to 5250 m a.s.l. in northern exposition.

Only the southeastern quarter of Nevado Coropuna has been investigated by Dornbusch (2002) using aerial photograph interpretation. Here, no active rock glaciers have been found but a number of very small fossil rock glaciers have been found with snout altitudes between 4650 m and 5000 m a.s.l.

On both Nevado Coropuna and Nevado Solimana, all the rock glaciers identified are well within the limit of previous glacial advances, and in particular the fossil rock glaciers occupy shaded positions beneath small vertical rock outcrops. These local conditions appear to have a higher influence on the snout distribution than general climatic variables. However, by comparing equilibrium line altitudes (ELA) for present and Pleistocene glacier extents, Dornbusch (2002) found that the Pleistocene ELA gradient was slightly lower than at present, indicating that the general circulation pattern was similar to that of today.

Cerro Yanahuara – Cerro Jollpa ridge (72°23′25″ W, 15°35′40″ S to 72°21′30″ W, 15°43′40″ S)

Cerro Yanahuara and Cerro Jollpa form the northern and southern end of a 16 km-long ridge

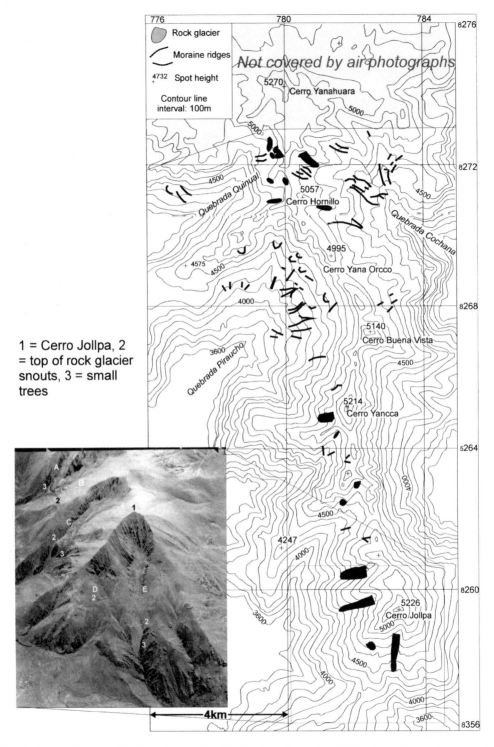

Fig. 2. Topographic map of the Cerro Jollpa Ridge (after Instituto Geographico Militar 1967). Contour line interval 100 m. Grid reference numbers refer to the UTM zone 19. Inset shows detail of aerial photograph no. 14064 of the HYCON flights in July 1955. Letters refer to features mentioned in the text

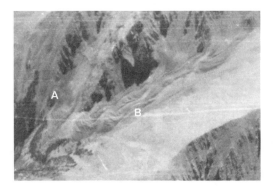

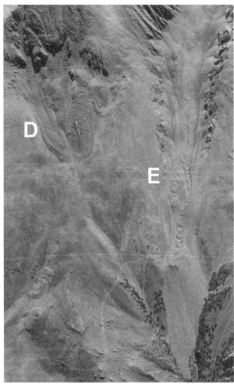

Fig. 3. Detail of the rock glaciers A, B, D and E mentioned in the text. For location see Fig. 2

(Figs 2 and 3) which exceeds 5200 m in altitude with several smaller summits. A detailed description of the glacial and periglacial features found along the ridge in Dornbusch (2002) is summarized below.

In the northern part of the ridge, around Cerro Yanahuara and Cerro Hornillo, the glacier–rock glacier assemblage found is very similar to that described previously for the Nevados Sara Sara, Solimana and Coropuna. A series of nested moraine ridges that can be subdivided, based on their altitudes and sizes, into different phases can be found in the lower parts of the valleys, while the higher parts are occupied by relatively small rock glaciers in shaded topographic locations favourable for rock glacier development. The rock glaciers appear to be inactive, although the highest ones may still be active, are small (less than 500 m long) and have an altitudinal range of less than 200 m. The upper part of the Quebrada Piraucho does not contain any rock glaciers at all except for those at Cerro Yancca. Here, moraine ridges become scarce and while moraines south of Cerro Hornillo reach below 4000 m a.s.l.; they

reach only down to about 4300–4500 m a.s.l. The rock glacier west of Cerro Yancca reaches down to 4700 m a.s.l. and might still be active.

At the southern end of the ridge at Cerro Jollpa, no moraines or other glacial features have been identified in the aerial photographs. Instead, the heads of the main valleys radiating south- and westward from Cerro Jollpa are occupied by long and narrow rock glaciers (Figs 2 and 3). They cover horizontal distances of up to 1 km and vertical distances of up to 400 m. Transverse surface ridges mainly occupy the lower parts towards the snouts, while the upper parts are generally smooth or show longitudinal ridges. The lower and middle parts also show 'trains' of transverse ridges that might indicate waves of renewed movement.

The most striking feature is the snout of rock glacier E, which extends from the top at *c.* 4400 to *c.* 4300 m a.s.l. The snout appears to be active, given its paler colour than the rock glacier surface, the pronounced break of slope between it and the surface, and the observation that it enters a tree-covered area with no indication of vegetation growing on the frontal slope.

In contrast, the much smaller rock glacier D to the west has its snout height at *c.* 4650 m a.s.l., and neither surface nor slopes show any signs of recent activity on the subdued topography.

A very similar situation to rock glacier E can be found at rock glacier C, where the snout reaches again to *c.* 4400 m a.s.l. and into an otherwise tree-covered area. However, the snout is much smaller and seems to lie on top of a fossil snout, which is densely covered with vegetation (between numbers 2 and 3). The upper part of D shows pronounced longitudinal ridges.

While rock glacier A seems to be inactive, rock glacier B again shows all signs of an active snout, that again enters tree-covered areas. Similar to C, rock glaciers A and B seem to lie on top of a fossil rock glacier whose the snout is outlined by the tree-covered up-slope form opening into a u-shaped form.

Discussion

No rock glaciers of comparable sizes or snout heights as those found at Cerro Jollpa have been found in the western part of the Cordillera Ampato. Their unique size might be related to the comparatively large but narrow cirques from which they originate. It might also be related to a steep Pleistocene precipitation gradient along the ridge, an idea which is supported by the decreasing number and increasing altitude of moraine features and corresponding ELA along the ridge towards the south. Geomorphological evidence from mountains like Cerro Achatayhua (Dornbusch 2000), west of the Nevado Sara Sara, point to the fact that small glaciers existed down to 4250 m a.s.l., indicating that along the Pacific edge of the western part of the Cordillera Ampato precipitation was high enough to facilitate the existence of glaciers rather than rock glaciers. This would suggest that, similar to today, regions of higher precipitation reached closer to the edge than further east.

The steep precipitation gradient along the Jollpa ridge is reflected in a rising trend of the ELA towards the south along the ridge (Fig. 4) that initially may have led to only small glaciers on the flanks of Cerro Jollpa. Owing to substantial rock walls in the valley heads, these glaciers could have become buried, leading to the formation of ice-cored rock glaciers. Seasonal snow, which still covered parts of Cerro Yancca and other parts of the ridge in the aerial photographs that were taken in July 1955 (the middle of the dry season), could then have contributed to the formation of interstitial ice that maintained the rock glacier activity. The slow response of rock glaciers to climatic changes due to the insulating properties of the surface cover of debris (Clark *et al.* 1994) may have kept these rock glaciers active for much of the Holocene to

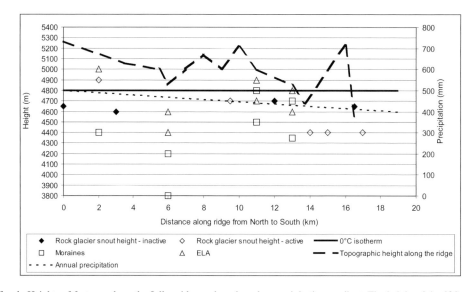

Fig. 4. Heights of features along the Jollpa ridge and modern day precipitation gradient. The height of the 0°C isotherm is based on Dornbusch 1998. The present day precipitation gradient shown is only tentative as the isohyets in Fig. 1 are interpolated and not based on observational data along the ridge

allow them to grow much larger and to reach much lower than other rock glaciers in the Cordillera Ampato.

These particular circumstances could also explain the apparent activity of the snouts in the 1950s despite their position at 4400 m a.s.l., about 500 m below the regional annual 0°C isotherm (Dornbusch 1998).

Only a field examination could verify the activity of the rock glaciers beyond doubt, but it can be speculated that the relatively well shaded position of the rock glaciers in the narrow valleys might have contributed to the survival of interstitial ice or an ice core over a long period of time through local microclimatic effects, and that cold phases such as the Little Ice Age might have reactivated movement that continued into the 1950s. Movement of rock glaciers that lie in well-shaded locations but altitudinally below the regional 0°C air temperature isotherm have also been reported from the European Alps (e.g., Vorndran 1969). Modelling of the ground surface temperature over the past 200 years by Guglielmin (2004) supports the view that rock glacier activity can still occur when the air-temperature would apparently preclude activity.

Conclusion

The changing assemblage of glacier and rock glacier features along the Cerro Jollpa ridge is unique in the Cordillera Amato and can be used to infer a steep local precipitation gradient during the late Quaternary. Although this study is only based on the interpretation of small-scale (c. 1:40,000) aerial photographs, it does show that large and active (as recent as the 1950s) rock glaciers exist at levels significantly below the regional annual 0°C isotherm. Only more recent aerial photographs, preferably at a larger scale, or field investigations can give details of present-day activity, and geophysical field investigations are necessary to support the interpretation on the genesis of the rock glaciers described.

The author would like to thank an anonymous referee for his useful comments and Brian Whalley for very helpful comments and suggestions for additional references to improve the manuscript.

References

ABELE, G. 1992. Landforms and climate on the western slope of the Andes. *Zeitschrift für Geomorphologie, Supplement*, **84**, 1–11.

AMES, A., DOLORES, S., VALVERDE, A., EVANGELISTA, P., CORCINO, D., GANVINI, W., ZÚÑIGA, J. & GOMEZ, V. 1988. Inventario de glaciares del Perú, Unidad de Glaciología e Hidrología, Huaraz, Peru.

BARSCH, D. & KING, L. 1989. Origin and geoelectrical resistivity of rockglaciers in semi-arid subtropical mountains (Andes of Mendoza, Argentinia). *Zeitschrift für Geomorphologie N.F.*, **33**, 151–163.

CLARK, D. H., CLARK, M. M. & GILLESPIE, A. R. 1994. Debris-covered glaciers in the Sierra Nevada, California, and their implications for snowline reconstructions. *Quaternary Research*, **41**(2), 139–153.

DOMARADZKI, J. 1951. Blockströme im Kanton Graubünden. *Ergebnisse der wissenschaftlichen Untersuchungen des schweizerischen National-parks*, **3**(24), 173–235.

DORNBUSCH, U. 1997. Geomorphological investigations on the late Quaternary glaciation in the western Andes of south Peru between 14°25′ S and 15°30′ S, inferable climatic conditions and their comparison with newly compiled data on the recent distribution of temperature, precipitation and snowline altitudes in south Peru south of 12°S. (title translated). PhD thesis.

DORNBUSCH, U. 1998. Current large-scale climatic conditions in south Peru and their influence on snowline altitudes. *Erdkunde*, **52**(1), 41–54.

DORNBUSCH, U. 2000. Pleistocene glaciation of the dry western Cordillera in southern Peru. Glacial Geology and Geomorphology; http://boris.qub.ac.uk/ggg/papers/full/2000/rp012000/rp01.html.

DORNBUSCH, U. 2002. LGM and present day snowline rise in the Cordillera Ampato, Western Cordillera, Southern Peru (15°15′–15°45′ S and 73°30′–72°15′ W). *Neues Jahrbuch für Geologie und Paläontologie*, **225**(1), 103–126.

FOX, A. N. 1993. Snowline altitude and climate in the central Andes (5–28°S) at present and during the late Pleistocene glacial maximum. PhD thesis, Cornell University, New York.

GROSJEAN, M., AMMAN, C., EGLI, W., GEYH, M. A., JENNY, B., KAMMER, K., KULL, C., SCHOTTERER, U. & VUILLE, M. 1996. Klimaforschung am Llullaillaco (Nordchile) – zwischen Pollenkörnern und globaler Zirkulation. *Jahrbuch der Geographischen Gesellschaft Bern*, **59**, 111–121.

GUGLIELMIN, M. 2004. Observations on permafrost ground thermal regimes from Antarctica and the Italian Alps, and their relevance to global climate change. *Global and Planetary Change*, **40**(1–2), 159–167.

HASTENRATH, S. L. 1967. Observations on the Snow Line in the Peruvian Andes. *Journal of Glaciology*, **6**, 541–550.

HEUSSER, C. J. 1989. Southern westerlies during the last glacial maximum. *Quaternary Research*, **31**, 423–425.

HOFFMANN, J. A. 1975. *Atlas climátologico de América des Sur*. WMO, UNESCO, Geneva.

INSTITUTO GEOGRAPHICO MILITAR. 1967. *Carta Nacional 1:100,000*, hoja 32r, Huambo.

KLEIN, A. G., SELTZER, G. O. & ISACKS, B. L. 1999. Modern and last glacial maximum snowlines in

the Central Andes of Peru, Bolivia and Northern
Chile. *Quaternary Science Reviews*, **18**, 63–84.

KRAINER, K. & MOSTLER, W. 2000. Reichenkar Rock
Glacier: a glacier derived debris–ice system in the
western Stubai Alps, Austria. *Permafrost and
Periglacial Processes*, **11**, 267–275.

MARKGRAF WRIGHT, H. E., KUTZBACH, J. E., WEBB
III, T., RUDDIMAN, W. F., STREET-PERROT, F. A.
& BARTLEIN, P. J. V. 1993. Climatic History of
central and south America since 18,000 yr b.p.:
comparison of pollen records and model simulations. *In*: WRIGHT, J. E. JR *et al.* (eds) *Global Climates since the Last Glacial Maximum*. University
of Minnesota Press, Minneapolis, MN, 357–385.

MARTIN, H. E. & WHALLEY, W. B. 1987. Rock glaciers.
Progress in physical Geography, **11**, 260–282.

MESSERLI, B., GROSJEAN, M. BONANI, G., Bürgi, A.,
GEYH, M. A., GRAF, K., RAMSEYER, K.,
ROMERO, H., SCHOTTERER, U., SCHREIER, H. &
VUILLE, M. 1993. Climate change and natural
resource dynamics of the Atacama Altiplano during
the last 18 000 years: a preliminary synthesis. *Mountain Research and Development*, **13**(2), 117–127.

NOGAMI, M. 1982. Ciculacío atmosférica durante la
última época glacial en los Andes. *Revist de
Geografía Norte Grande*, **9**, 41–48.

SILVA, S. L. DE & FRANCIS P. W. 1991. *Volcanoes of
the Central Andes.* Springer, Berlin.

VORNDRAN, E. 1969. Untersuchungen über Schuttent-
stehung und Ablagerungsformen in der Hochregion
der Silvretta (Ostalpen). *Schriften des Geographischen Instituts der Universität Kiel*, **29**(3), 138.

WAHRHAFTIG, C. & COX, A. 1959. Rock glaciers in
the Alaska Range. *Bulletin of the Geological
Society of America*, **70**, 383–436.

WEBERBAUER, A. 1922. Die Vegetationskarte der
peruanischen Anden zwischen 5° und 17°S. *Peter-
manns Geographische Mitteilungen*, **68**, 89–91.

WHALLEY, W. B. & AZIZI, F. 2003. Rock glaciers and
protalus landforms: analogous forms and ice
sources on Earth and Mars. *Journal of Geophysical
Research*, **108**(E4).

Cryological processes implied in Arctic proglacial stream sediment dynamics using principal components analysis and regression

T. D. L. IRVINE-FYNN[1,5], B. J. MOORMAN[2], D. B. SJOGREN[2], F. S. A. WALTER[1],
I. C. WILLIS[3], A. J. HODSON[4], J. L. M. WILLIAMS[1] & P. N. MUMFORD[4]

[1]*Department of Geography and* [2]*Earth Science Program, University of Calgary,*
2500 University Drive N.W., Calgary, Alberta, Canada T2N 1N4,
[3]*Scott Polar Research Institute, University of Cambridge, Lensfield Road,*
Cambridge CB2 1ER, UK
[4]*Department of Geography, University of Sheffield, Winter*
Street, Sheffield S10 2TN, UK
[5]*Present address: Department of Geography, University of Sheffield,*
Winter Street, Sheffield S10 2TN (email: t.irvine-fynn@sheffield.ac.uk)

Abstract: In high latitudes, recent research has demonstrated that both thermo-erosion and temperature dependence influence sediment release into fluvial systems. An analysis of proglacial suspended sediment concentration (SSC) dynamics is presented for three glacierized basins: cold-based Austre Brøggerbreen (Svalbard), polythermal Midre Lovénbreen (Svalbard) and polythermal Glacier B28 (Bylot Island). The temporal variation in processes dominating SSC patterns is assessed using stepwise multivariate regression following the subdivision of the time series. Partitioning of the time series is achieved through principal components and change point analyses. The regression models use discharge and surrogate predictor variables to model SSC, while improvements are made by using air temperature and radiation terms as independent variables. Comparisons are drawn between two sets of models with contrasting subseasonal division. By interpretation of the regression model characteristics, temporal changes in physical processes are implied over the course of the time periods. Numerical analyses suggest there is a trend for changes between fluvial, glacial and periglacial factors forcing responses in SSC. Therefore, it is conjectured that glaciofluvial sediment transfer at high latitudes is influenced by periglacial processes and conditions. This has implications for the predictions of fluvial sediment loads in a changing environment, and the use of sedimentary records for environmental reconstruction.

In the Cascade Mountains, British Columbia, Evans (1997) concluded that spatio-temporal climatic variability was a fundamental control on glaciofluvial sediment yield. Highest sediment yields corresponded to cool, moist conditions typical of deglaciation periods (Evans 1997); this implies that subaerial periglacial processes and ice degradation may supply significant volumes of sediment. A recent model presented by Morehead *et al.* (2003) suggests that sediment load may indeed be partially dependent on basin temperature at all latitudes. Despite relatively low sediment loads discharged by Arctic rivers, it has been predicted that a 2°C rise in atmospheric temperature will lead to a 22% increase in fluvial sediment load (Syvitski 2002). Currently, in high latitudes, release of unconsolidated detritus from glacial, proglacial and recently deglaciated Neoglacial valley source areas is the principal origin of sediment delivered to the Arctic oceans. However, little attention has been paid directly to the potential influence of the ice-proximal environment in sediment delivery to Arctic proglacial streams.

In both the Eurasian and Canadian Arctic, massive ground ice has been observed in deglaciated and ice-proximal terrain. It is suggested that much of the massive ground ice may in fact be buried glacier ice, protected from melting by a debris cover in excess of the active-layer thickness (Lorrain & Demeur 1985; Kaplanskaya & Tarnogradskiy 1986; Astakhov & Isayeva 1988; French & Harry 1990; Astakhov 1992; Evans & England 1993; Robinson *et al.* 1992; Vaikmäe *et al.* 1993; Etzelmüller *et al.* 1996; Moorman & Michel 2000a). Degradation of these massive

From: HARRIS, C. & MURTON, J. B. (eds) 2005. *Cryospheric Systems: Glaciers and Permafrost.*
Geological Society, London, Special Publications, **242**, 83–98.
0305-8719/05/$15.00   © The Geological Society of London 2005.

ground ice forms occurs by thermo-erosion induced by changing regional or local climatic conditions.

In Svalbard, changes in the surface elevation of five proglacial areas over the course of two decades were interpreted as thermo-erosion of ice-cored moraine forms in the ice-proximal zone (Etzelmüller 2000). Holmlund *et al.* (1996) conjectured that both downvalley movement of ice-cored moraine forms and spatially discontinuous melting of permafrost occur in the proximal zone of Storglaciären, northern Sweden. Such results indicate thermo-erosion of glaciated and deglaciated terrain might rapidly contribute large quantities of sediment available for transport. Thermo-erosion may be intimately coupled with glaciofluvial erosion, deposition and proglacial channel migration, thereby accentuating sediment exchange, release and storage within short distances from a glacier margin.

Research focussed on Arctic proglacial suspended sediment transport remains underrepresented and contrasting (e.g. Bogen 1991; Vatne *et al.* 1992; Gurnell *et al.* 1994; Sollid *et al.* 1994; Hodgkins 1996, 1999; Hodson *et al.* 1997, 1998a; Hodson & Ferguson 1999). The contrasts may be explained by differences in glacier thermal regime, temporal characteristics, mass balance and meteorological conditions. Typically, however, the suspended sediment concentration (SSC) dynamics demonstrate a tendency for short-term autocorrelation, even within regression model residuals (Gurnell *et al.* 1994; Hodgkins 1999; Hodson & Ferguson 1999). Such results suggest stochastic, transient sediment flushes, independent of stream discharge, passing though the glaciofluvial system. Unexplained variability in Arctic stream SSC has been argued, qualitatively, to evoke variable ground thaw processes (Gurnell *et al.* 1994), ground thaw and solifluction (Hodson *et al.* 1997, 1998a), and the potential influence of denudation of ice-cored moraines (Hodson & Ferguson 1999). However, to date, researchers examining glaciofluvial processes have placed little direct emphasis on periglacial processes including thermo-erosion, thaw-induced debris flows, mass wasting and debris slumps, and solifluction in ice-marginal areas.

This paper presents a variety of regression models that examine the significance of thaw-related processes of sediment supply. Proglacial stream data and corresponding meteorological data from southeast Bylot Island (Nunavut) and northeast Spitsbergen (Svalbard) are subdivided using principal components analysis (PCA) and change point analysis (CPA). Subseasonal regression analysis enables temporal changes in the dominance of glacial, fluvial and periglacial processes to be assessed.

Field site characteristics

Midre Lovénbreen and Austre Brøggerbreen (Fig. 1) are two small, north-facing valley glaciers located on the Brøggerhalvøya Peninsula, western Spitsbergen, Svalbard (78°50′ N, 12°00′ E). Continuous permafrost in the area ranges in depths from 100 to 400 m (Liestøl 1977). The geology of the peninsula is complex, but Caledonian age metamorphic rocks (phyllite and schists) exist virtually throughout the glacierized basins, with some overlying sedimentary rocks (sandstones and carbonates) occurring, more so at Austre Brøggerbreen (Hjelle 1993; Hodson *et al.* 2000). The glaciers extend from *c.* 50 m to *c.* 600 m above sea level (a.s.l.), both being of maximum thicknesses <180 m with an ELA *c.* 400 m a.s.l. Both glaciers have been characterized by negative mass balances since the end of the Little Ice Age (Hagen & Liestøl 1990; Lefauconnier & Hagen 1990).

The 6.1 km^2 Austre Brøggerbreen catchment is now thought to be entirely cold-based (cf. Hodson *et al.* 1998b; Hodson & Ferguson 1999). As a result, meltwater drainage is primarily routed sub-aerially in marginal and supraglacial channels. There is a well-developed englacial system (Hagen *et al.* 1991; Hodson *et al.* 2002) that appears to emerge close to the glacier surface on the eastern margin. Midre Lovénbreen, covering an area of 5.5 km^2, is polythermal, with a temperate ice core and cold margins (Björnsson *et al.* 1996). The temperate ice zone is in the accumulation and upper ablation areas (Björnsson *et al.* 1996; Rippin *et al.* 2003). Summer meltwater is only partly drained supraglacially, with a subglacial system discharging significant quantities of water to the proglacial region. Annually, this subglacial system emerges at the snout of the glacier as a turbid artesian upwelling (Hodson *et al.* 2000), indicating the temporary storage and enhanced water pressure at the glacier bed prior to the hydrological breach of the cold margin.

The south-facing Glacier B28 (Inland Waters Branch 1969), unofficially named 'Stagnation Glacier' (72°58′ N 78°22′ W), is located on the south side of Bylot Island, Nunavut (Fig. 1). Moorman and Michel (2000b) suggested that the 14.0 km^2 Stagnation Glacier is polythermal. Aerial imagery between 1948 and 1994 suggests Stagnation Glacier has been retreating at *c.* 26 m/a, while observations since 1999 indicate that retreat has slowed to approximately 5 m/a. The glacier snout is at *c.* 320 m a.s.l., and the ELA is at *c.* 1050 m a.s.l. The basin's geology is primarily Archean igneous and metamorphic

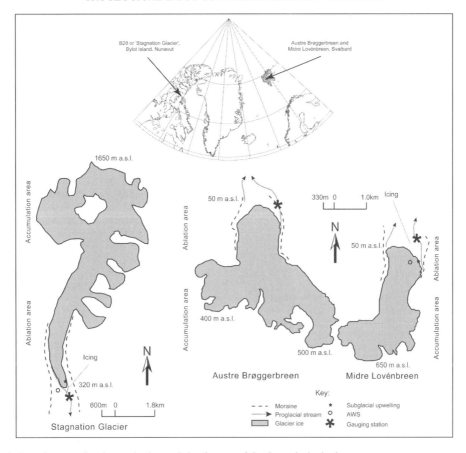

Fig. 1. Location map showing study sites and sketch maps of the three glacier basins

rocks (Jackson *et al.* 1975), and local continuous permafrost depths reach a maximum of *c.* 400 m (Moorman & Michel, 2000a). Subaerial hydrology dominates Stagnation Glacier; marginal channels and deeply incised supraglacial channels are the principal drainage routes. Temporary vertical displacement of the glacier was recorded in 2000, possibly indicative of transient subglacial water storage. Dye trace experiments in 2001 and 2002 suggest the existence of en- and/or subglacial drainage.

Because Austre Brøggerbreen is a cold-based glacier, no 'winter runoff' occurs. At the two polythermal glaciers, Midre Lovénbreen and Stagnation Glacier, winter runoff enables naled icing forms to develop. At both glaciers, icings extend up to 500 m from the snout, occupying proglacial topographic depressions.

Methods

In each glacierized basin a proglacial gauging station was emplaced approximately 500 m from the glacier margin. The sites were selected because they had relatively stable upstream channel reaches, with minimal channel migration or braiding. This minimized sediment contributions from channel migration and modification. The two Svalbard basins were monitored in 2000, and Stagnation Glacier in 2002. Continuous records of discharge (*Q*) and suspended sediment concentration (SSC) were collected. In-stream pressure transducer data were corrected through the use of calibration curves derived from velocity-area estimates of *Q*.

Determination of SSC involved calibration curves obtained from filtered water samples. The frequency of sampling was four times per day throughout the monitoring period, and SSC samples (with errors of *c.* 2%) were plotted against corresponding hourly-averaged turbidity data. Infrared turbidity probes were used to minimise the factors which may compromise data (see Clifford *et al.* 1995). Resulting primarily from the forecasting uncertainty of the rating curves, typical errors in all hydrological time series were calculated from the standard error of the regression between measured and modelled data points. For Q, the uncertainty was between 15 and 28%, while errors in SSC were noted to be up to 68% reflective of the short-term temporal variability in sediment transport.

Complementary data collection included continuous records of mean hourly air temperature (AT), incoming radiation (IR) and net radiation (NR). Automatic weather stations (AWS) were located on Midre Lovénbreen's snout (*c.* 110 m a.s.l.) and on a moraine at the margin of Stagnation Glacier (*c.* 350 m a.s.l.). We assume, due to the adjacent positioning of Midre Lovénbreen and Austre Brøggerbreen, that the one AWS is representative for both basins.

Results and initial interpretations

Figure 2 presents annotated graphs of the Q and SSC data for each glacier basin. Occasional gaps

($<$6 h) in data series due to short-term instrument failure occupied $<$3% of any data set. The gaps were 'filled' using a recursive Kalman random walk filter, ideally suited to non-stationary data (Young *et al.* 1991; Hodson *et al.* 1998b).

The similarity in the pattern of Q at Austre Brøggerbreen and Midre Lovénbreen is expected for two neighbouring, geometrically similar glaciers over the same ablation season. Despite the similar Q, large disparities are seen in the SSC record. The sudden increase in SSC at Midre Lovénbreen on Julian Day (JD) 198 corresponded to the upwelling of turbid subglacial meltwater, emerging on the eastern side of the glacier snout. At this time, SSC jumped from a mean of 0.27 g/l prior to midday JD198, to a mean of 1.78 g/l for the remainder of the season. For the latter half of the season SSC declined at Midre Lovénbreen, suggesting that exhaustion of the subglacial reservoir occurred.

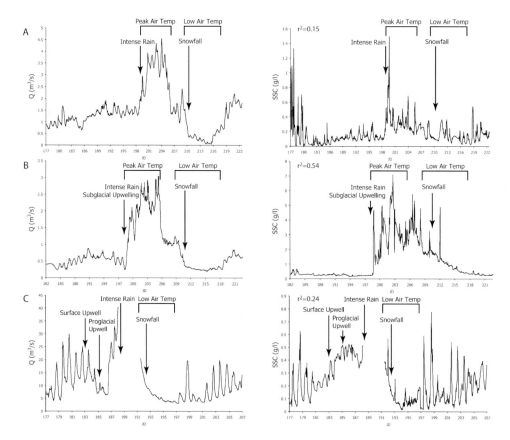

Fig. 2. Graphs of raw SSC and Q data. (**A**) Austre Brøggerbreen; (**B**) Midre Lovénbreen; and (**C**) Stagnation Glacier. Note, the axes scales differ for each glacier. Coefficient of determination between SSC and Q is shown

Austre Brøggerbreen also demonstrated a brief peaked rise in SSC on JD198–199. The upwelling at the polythermal glacier, and peak in the cold-based basin coincided with a rain event. The peaks in SSC, early in the season at Austre Brøggerbreen, are thought to indicate the initiation of the channel at the beginning of the melt season, and are not related to Q.

At Stagnation Glacier, on JD183 a turbid upwelling was observed on the glacier surface approximately 1.9 km upglacier from the snout. This upwelling ceased on JD185, and a turbid outflow at the glacier margin was noted. Combined with the record of surface uplift in 2000, we suggest this is evidence of a subglacial reservoir. The supraglacial upwelling occurred as subglacial water pressures forced meltwater up through fractures or weaknesses in the ice. These weaknesses may be associated with a crevasse zone c. 100 m further upglacier from the upwelling site. The supraglacial upwelling ceased once an en- or subglacial drainage route was established, passing directly to the snout. The break in the series for JD189–192 corresponds to a rainstorm which caused flooding of the proglacial region necessitating the removal of data logging equipment.

Examination of all the SSC series revealed that SSC peaks did not always coincide with the diurnal Q maxima or rainfall events. Furthermore, maximum rates of change in Q did not appear to be a forcing mechanism for SSC. Such results indicate alternative sediment supply mechanisms since changes in SSC were not simple responses to variations in Q.

Statistical analysis

The following analyses utilize an objective methodology to subdivide the time series data, and then examine differences between forcing mechanisms acting on SSC. Process-based inferences can be made from the use of multiple proxy variables for each subseasonal period.

One of the limitations of many SSC studies is the examination of the entire season as a whole. Commonly, time-series data have been divided on the basis of observed temporal changes in meteorological or hydrological variables (e.g. Gurnell *et al.* 1992, 1994; Hodson *et al.* 1998b). Hannah *et al.* (1999) used one alternative approach by cluster analysis of like-hydrographs, while Richards and Moore (2003) used a cumulative sum (CUSUM) technique, dividing a stream discharge time-series by visual determination of changes in the slope of the CUSUM curve. However, using these approaches, there

is subjectivity in both the definition and location of the boundaries between subseasons. For this reason, we opt for a more objective methodology using PCA and CPA. This method can thus incorporate both meteorological and hydrological data, and statistically divide the time-series into subperiods.

Principal components analysis

PCA has been shown to be an effective statistical tool particularly applied to water chemistry (e.g., Nolan 1999; Sánchez-Martos *et al.* 2001; Petersen *et al.* 2001; Haag & Westrich 2002; Hodson *et al.* 2002), and yet has been used infrequently in glaciofluvial analyses. Analysis is accomplished by extracting principal components (PCs) or linear combinations of the original variables, defined by the variance–covariance (or correlation) matrix. PCs are uncorrelated and therefore represent independent 'modes' or 'dimensions' of variance within the multivariate data. Subsequent components explain progressively less of the total variance. The more effective analyses are obtained when original variables are correlated (Table 1).

Linking PCA results to physical processes, or reification, is a subjective process (Davis 1986; Haag and Westrich 2002). This is done by interpreting the component 'scores' and 'loadings'. Loadings represent the correlation coefficient between the original variable and the PC. Scores for any one component are created by projecting the original data using the coefficients defined by the PC (i.e., the standardized loading which is the PC loading divided by the eigen-

Table 1. *Correlation matrix for all variables*

Glacier	Variables	SSC	Q	AT	NR	IR
AB	SSC	1				
	Q	0.46	1			
	AT	0.45	0.60	1		
	NR	0.21	0.28	0.43	1	
	IR	−0.06	−0.29	0.11	0.57	1
ML	SSC	1				
	Q	0.38	1			
	AT	0.07	0.62	1		
	NR	−0.05	0.31	0.48	1	
	IR	0.01	−0.15	0.04	0.65	1
SG	SSC	1				
	Q	0.62	1			
	AT	0.47	0.57	1		
	NR	0.22	0.25	0.39	1	
	IR	0.17	0.22	0.40	0.98	1

Note the strong correlation occuring between IR and NR at the Stagnation Glacier Basin. Glacier names are abbreviated.

value). Loadings with opposing signs indicate an inverse relationship between the two original variables, possibly indicating a lagged relationship for out-of-phase variables.

A PCA was applied to all data collected (Q, SSC, AT, IR and NR). The radiation terms are considered proxies for albedo (snowline position) and cloud cover (synoptic weather patterns). Precipitation data were not included; permafrost ensures Q and precipitation are interdependent, being closely linked due to the rapidity of runoff. The results of the PCA are presented in Table 2. The first component (PC1) explains *c.* 50% of the data variance. No standardized loading is particularly high (all being <0.5). Therefore, PC1 describes how all variables co-vary over time, indicative of the seasonal pattern of change of both meteorological and hydrological variation. This, we argue, includes variations in physical processes and synoptic weather patterns which have been documented to influence hydrological conditions (e.g., Hodson *et al.* 1998b).

PC2 varies for each basin, but at Stagnation Glacier the standardized loadings for Q and SSC dominate (both *c.* 0.5). The positive sign for both variables indicates that a positive relationship between SSC and Q is the dominant factor, explaining 26% of the data's variance. For the two Svalbard basins, IR and NR dominate PC2. With IR in both cases exhibiting a loading value >0.5, the interpretation is that diurnal variability in radiation in the Svalbard basins

explains approximately 30% of the total variance. PC3 demonstrates the relationships for SSC in all basins. Examining the loading values, SSC and Q are inversely related at Stagnation Glacier, explaining a further 10% of the data variability; at Midre Lovénbreen, AT and SSC appear inversely linked; and the independence of SSC variation dominates PC3 at Austre Brøggerbreen. PC4 and PC5 explain only *c.* 10% of the variance cumulatively in all three basins. However, these two latter PCs can effectively be considered 'noise' and are difficult to interpret in physical terms.

The analysis shows that differences exist between glacier basins; in particular the links between meteorology and hydrology are varied. However, importantly, the relationships of seasonal trend (PC1) and SSC variability not positively relating to Q (PC3) potentially enable subdivision of the time-series based on those physical relationships.

Change point analysis

CPA is a powerful tool for determining the dynamics of a time series. The statistical method is capable of detecting multiple, subtle changes and providing confidence in those change points (Taylor 2000). The approach is based on the analysis of the CUSUM plot (as used by Richards & Moore 2003). The CUSUM at time t for a time

Table 2. *Results of PCA extracting five principal components*

Glacier	Variable	PC1	PC2	PC3	PC4	PC5
AB	Q	0.330	−0.354	−0.405	−0.260	1.445
	SSC	0.266	−0.195	1.045	−0.125	−0.027
	AT	0.384	−0.107	−0.301	0.926	−1.134
	IR	0.149	0.605	0.159	0.698	1.040
	NR	0.304	0.392	−0.173	−1.121	−0.622
	% σ^2	45.3	28.8	14.3	7.4	4.1
ML	Q	0.351	−0.282	0.021	−0.99	1.677
	SSC	0.285	−0.306	0.780	0.626	−1.173
	AT	0.339	−0.028	−0.846	1.093	−0.218
	IR	0.154	0.515	0.476	0.819	1.296
	NR	0.282	0.413	−0.058	−1.162	−1.352
	% σ^2	47.4	32.8	12.9	4.4	2.6
SG	Q	0.208	0.475	1.055	0.208	0.169
	SSC	0.215	0.463	−0.885	0.640	0.097
	AT	0.291	0.151	−0.252	−1.222	−0.164
	IR	0.318	−0.374	0.084	0.308	−4.901
	NR	0.314	−0.392	0.056	0.246	4.929
	% σ^2	53.5	26.5	10.2	9.5	0.4

Standardized loadings for each variable in sequential components are shown. The percentage of variance (% σ^2) explained by each individual PC is also indicated. Glacier names are abbreviated.

series (C_t) is given as:

$$C_t = \sum_{t=1}^{t}(x_i - \mu) \qquad (1)$$

where x_i is the time-series variable under investigation and μ is the series mean.

Changes in the direction and slope of the CUSUM plot indicate changes in the data trend. However, to determine these changes objectively, a statistical method called bootstrapping is employed. Bootstrapping involves the random reorganisation of x_i values, from which new CUSUM curves are calculated. The idea behind bootstrapping is that the bootstrap CUSUM curves represent data that mimic the behaviour of the original CUSUM if no change has occurred (Taylor 2000). The difference between the maximum and minimum CUSUM values (S_{diff}) of each bootstrap is compared with the original data's S_{diff} to determine if this value is consistent with what would be expected if no change occurred (Taylor 2000). In this manner, change points can be identified, and assigned a confidence of change. Once a change point has been detected, the data are broken into two segments and the analysis repeated for each segment (Taylor 2000). Chen & Gupta (2000) and Taylor (2000) provide comprehensive reviews of the technique.

To identify subseasons with characteristically different hydro-meteorological conditions, we perform a CPA on the score series from the PCA. PC1 and PC3 are used to examine seasonal trends and patterns where SSC is not related directly to Q. Departures from the relationships given by each PC are seen in the score series as deviations away from the standardized mean, which is approximately zero. The PC standardized score trends for PC1 and PC3 in all three basins indicate differing periods by data trending away from the mean and drifts returning towards the zero (Fig. 3).

CPA and bootstrapping rest on a single assumption, that of an independent error structure (Taylor 2000). Where errors are autocorrelated or positively correlated, typical in strongly cyclical data, the analysis may indicate change points incorrectly. To resolve this issue we confirmed the presence of a diurnal cycle in all score time-series through the use of Fourier Analysis. Therefore, we assume that data points spread 24 h apart are sufficient to identify breaks within the data set as a whole. However, this ensures our change points can only be given as days rather than specific hours. Scores for 00.00 (midnight) were used in the CPA, which was conducted using Change Point Analyser software, performing 1000 bootstraps for each CPA procedure. The change points dividing the data for each PC score time series are indicated in Fig. 3, with confidence in the change indicated. While some subseasons for PC1 and PC3 coincide, there are subtle differences, and a tendency for fewer subseasons to be identified by analysis of PC3. The rationale for this difference is that PC3 represents a purely hydrological series, focused upon the sediment dynamics, while PC1 embodies the seasonal trends in weather, snow cover and glacio-fluvial character. To examine the processes contributing to SSC dynamism, we follow a multivariate regression procedure.

Multivariate regression

The coefficients of determination resulting from simple bivariate regression between Q and SSC were low, being less than 0.55 (see Fig. 2). Researchers have shown that changes in sediment supply can be incorporated into regression models using proxy variables to examine sediment transport processes not simply linked to instantaneous discharge values (e.g. Richards 1984; Willis et al. 1996; Hodgkins 1999; Hodson & Ferguson 1999). Multivariate regression models (MRMs) are thus able to give indication of the processes contributing to the dynamic changes in stream sediment load.

A $\log_{10}$ transformation was applied to the raw SSC time-series because it is necessary to normalize the dependent variable when using MRMs. Similarly, a $\log_{10}$ transformation was applied to the raw Q data creating a 'log$Q$' variable to indicate instantaneous forcing of 'logSSC'. To account for diurnal hysteresis, the rate of change in log$Q$ (hereafter δQ) was calculated and used as an independent variable. Positive regression coefficients for δQ indicate clockwise hysteresis where SSC is highest with rising discharges, and negative coefficients for δQ indicate counter-clockwise hysteresis. A cumulative sum of discharge (ΣQ) is used to indicate sediment supply or exhaustion. Here, ΣQ is given by addition of instantaneous Q estimates rather than actual cumulative Q for each hour. Positive regression coefficients between logSSC and ΣQ indicate increasing sediment supply, while negative coefficients indicate sediment exhaustion despite the increasing flux of meltwater passing though the drainage system. A Q^2 variable was used to examine non-linearity (e.g., Richards and Moore 2003). A positive coefficient for Q^2 implies that small increases in Q lead to larger nonlinear increases in SSC. This, we argue, relates to stream flow and hydraulic radius, demonstrating, for this high-latitude

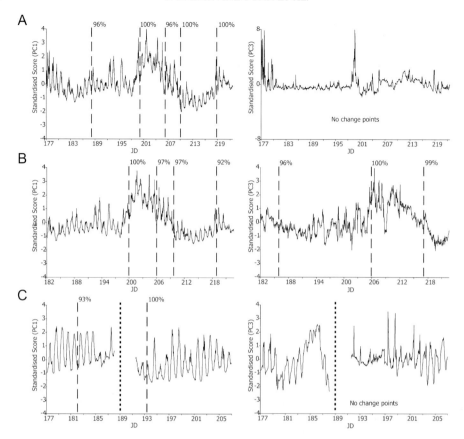

Fig. 3. Graphic of standardized scores for PC1 and PC3 for each basin. (**A**) Austre Brøggerbreen; (**B**) Midre Lovénbreen; and (**C**) Stagnation Glacier. Note, the axes scales differ for each glacier. Change points are indicated by dashed lines, and the corresponding confidence is shown in percent. The break in Stagnation Glacier's data is indicated with a broad dotted line

environment, the sensitivity of permafrost-influenced stream margins. Destabilization of permafrost-influenced sediments may occur in response to thermo-erosion initiated by contact with meltwater streams at the banks and bed. Entrainment is thus sensitive to small increases in hydraulic radius, stream turbulence and viscous dissipation both at the bed and unconsolidated banks. The variable HQ is defined as the time since Q at time t, was last equalled or exceeded in the time-series. A precipitation variable (Richards 1984) was not included since rainfall and Q are not truly independent and Arctic rainfall tends to be low in magnitude (Willis *et al.* 1996; Hodgkins 1999), while synoptic weather patterns and radiation budgets are closely related to glacial discharge (Hodson *et al.* 1998b). Predictor variables were initialized at the start of each subperiod.

To supplement a purely hydrologically defined MRM, energy terms (AT, IR and NR) were used to examine potential sediment supply processes directly relating to thermo-erosional release at the ice margin. The correlation matrix (Table 1) shows that in the Svalbard basins, the variables are relatively weakly correlated ($r < 0.8$), despite some being significant at the 95% confidence level. However, at Stagnation Glacier, IR and NR are strongly correlated ($r > 0.9$). We therefore use only AT and NR to give indication of periglacial supply processes occurring in the ice-proximal environment for Stagnation Glacier, and all three variables for Austre Brøggerbreen and Midre Lovénbreen.

Stepwise linear MRMs, were developed for each subseason identified through the use of the PCA and CPA, with entry F-statistics of 0.05. Figure 4 presents the regression results in graphi-

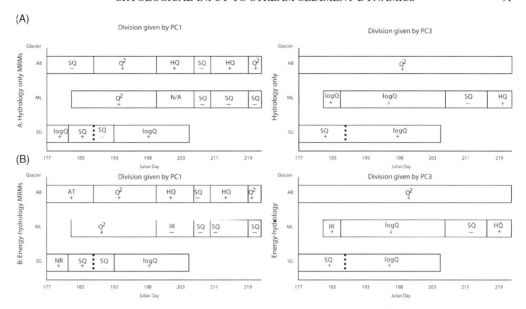

Fig. 4. Schematic illustrating the more significant predictors in modelling SSC for (**A**) hydrology only models, and (**B**) energy–hydrology models for subseasonal division based on PC1 and PC3. The sign of the predictor coefficient is indicated. A hydrology model could not be defined for ML3-PC1

cal format; the most significant variable in each MRM is indicated, along with the sign of the coefficient. Table 3 presents the model statistics, and clearly shows incorporation of energy terms can increase the coefficient of determination (R^2) by up to 0.3, although for the vast majority of the models, increases of less than 0.1 are achieved. Notably, model improvements using energy predictors are seen predominantly for periods early in the season at all three glaciers.

Autocorrelation

Examination of the coefficients of determination (Table 3) shows that the MRMs are only partially successful in prediction of SSC. Analysis of the MRM residuals followed a simple examination of plots of the autocorrelation and partial autocorrelation functions (ACF and PACF, respectively). All model residuals showed significant yet declining ACF at short lags (<6 h), with a corresponding significant PACF occurring at a lag of 1 h. Such a pattern is suggestive of a non-seasonal autoregressive process, i.e., SSC at time t is partially dependent on SSC at ($t - 1$) (see Vandaele 1983). Autocorrelation in multivariate regression residuals was found by Willis *et al.* (1996) and Hodson & Ferguson (1999). Willis *et al.* (1996) applied an autoregressive integrated moving average (ARIMA) model to residuals, while Hodson and Ferguson

(1999) included a lagged SSC predictor in regression models. Willis *et al.* (1996) interpreted an autoregressive ARIMA model to imply that SSC was dependent on previous values. However, Hodson and Ferguson (1999) suggested that physical interpretation of ARIMA models on residual series is not simple. Nonetheless, because the lagged SSC variable ($SSClag_n$) is not truly independent of Q and is strongly correlated to unlagged SSC, Hodson and Ferguson (1999) found that '$SSClag_n$' dominated regression models.

To reduce residual autocorrelation and to avoid the strong correlation between SSC and a lagged SSC variable, we use a lagged residual variable as applied by Willis *et al.* (1996). Because ACF and PACF charts suggested an autoregressive processes is extant in the data sets, the residual series from the hydrology-energy models lagged by 1 h (R_{t-1}) was included in a final set of MRMs. The coefficients of determination are shown in Table 4, and demonstrate that significant improvement is achieved by recognition of autocorrelation in the SSC series. This suggests SSC is strongly responsive to stochastic supply processes, and the temporarily elevated sediment availability takes time to become exhausted. In this case, time for sediment slugs to pass through the glaciofluvial system is shown to be of the order of 1 h or more. Plots of ACF and PACF for the residuals

Table 3. *Results from the multivariate regression*

MRM	R^2 (HO)	R^2 (HE)	Increase in R^2	RSS (HO)	RSS (HE)	Total RSS (HO)	Total RSS (HE)
AB1-PC1	0.243	0.412	0.17	47.63	35.94		
AB2-PC1	0.668	0.775	0.11	5.53	3.70		
AB3-PC1	0.250	0.250	0.00	1.90	1.90		
AB4-PC1	0.756	0.815	0.06	0.70	0.52	66.38	51.72
AB5-PC1	0.417	0.463	0.05	9.36	8.53		
AB6-PC1	0.582	0.620	0.04	1.26	1.13		
AB1-PC3	0.251	0.316	0.01	125.31	114.16	125.31	114.16
ML1-PC1	0.588	0.605	0.02	9.86	9.38		
ML2-PC1	N/A	0.311	0.31	N/A	2.00		
ML3-PC1	0.848	0.848	0.00	0.30	0.30	13.47*	12.93
ML4-PC1	0.919	0.923	0.00	1.15	1.09		
ML5-PC1	0.346	0.346	0.00	0.16	0.16		
ML1-PC3	0.290	0.566	0.28	0.86	0.52		
ML2-PC3	0.840	0.861	0.02	21.50	18.60		
ML3-PC3	0.857	0.880	0.02	2.96	2.46	25.38	21.64
ML4-PC3	0.590	0.614	0.02	0.06	0.06		
SG1-PC1	0.460	0.707	0.25	4.49	2.39		
SG2-PC1	0.820	0.820	0.00	0.31	0.31		
SG3-PC1	0.903	0.903	0.00	0.14	0.14	21.30	18.04
SG4-PC1	0.572	0.602	0.03	16.36	15.20		
SG1-PC3	0.643	0.718	0.08	7.08	5.57		
SG2-PC3	0.546	0.546	0.00	18.77	18.77	25.85	24.34

Coefficients of determination are compared between hydrology-only (HO) and hydrology–energy (HE) models, with increases shown. Note the abbreviation of subseasons (i.e., AB1-PC1 refers to Austre Brøggerbreen subseason 1 defined by PC1). Residual sum of squares (RSS) are also shown with total season RSS for each glacier for models based on PC1 and PC3, respectively. The asterisk represents spurious values for Midre Lovénbreen where no model could be identified.

from the new MRMs demonstrated that the majority of autocorrelation was removed; however in some cases singular significant autocorrelation remained at a lag >2 h. For only five of the MRMs did R_{t-1} usurp the predictors identified in Fig. 4: in AB3-PC1, AB5-PC1, AB1-PC3, ML2-PC1 and ML5-PC1. For these periods we suggest that patterns of SSC are dominated by self-stimulating or autogenic processes, and not hydrologically forced by discharge or energy conditions. The implication of this is that processes of sediment entrainment, channel migration, solifluction, bank collapse, debris flows and the like occur indirectly linked to hydrological characteristics.

Discussion

Subseasonal division

In broad terms, the CPA subdivisions based on PC1 (Fig. 3) correspond well to observed meteorological conditions. Unsurprisingly, the divisions for Austre Brøggerbreen and Midre Lovénbreen are similar. The break around JD199 can be linked to increased air temperatures and a switch from snow melt- to ice melt-dominated conditions on the glacier, following the removal of snow at lower elevations in response to an intense rainfall event. The subseasonal divide at JD206 coincides with a reduction in air temperatures, following the seasonal maximums, and increased cloud cover. This period was followed by snowfall on JD209–210. The snowline returned to lower elevations on the glacier tongue from this time until JD218 when renewed, effective snowmelt occurred. These processes are picked up well by the CPA. The division earlier in the season at Austre Brøggerbreen is interpreted to link to the pattern of declining air temperatures until JD184 and rising thereafter.

At Stagnation Glacier, the series is artificially broken by the forced removal of logging equipment on JD189. However, rising air temperatures characterized JD177–181, with consistent patterns continuing from JD182 to 188, apparently corresponding to the divide identified by the CPA. Following the storm, snow fall brought the snowline back to the glacier snout from JD191 to 193, with low air temperatures and decreased diurnal variation in incoming and net

Table 4. *Results comparing multivariate models accounting for autocorrelation*

MRM	R^2 (HE)	R^2 (HE $+ R_{t-1}$)	Increase in R^2
AB1-PC1	0.412	0.659	0.25
AB2-PC1	0.775	0.901	0.13
AB3-PC1	0.250	0.684*	0.43
AB4-PC1	0.815	0.877	0.06
AB5-PC1	0.463	0.861*	0.40
AB6-PC1	0.620	0.783	0.16
AB1-PC3	0.316	0.797*	0.48
ML1-PC1	0.605	0.886	0.28
ML2-PC1	0.311	0.657*	0.35
ML3-PC1	0.848	0.884	0.04
ML4-PC1	0.923	0.965	0.04
ML5-PC1	0.346	0.754*	0.41
ML1-PC3	0.566	0.777	0.21
ML2-PC3	0.861	0.981	0.12
ML3-PC3	0.880	0.955	0.08
ML4-PC3	0.614	0.779	0.17
SG1-PC1	0.707	0.870	0.16
SG2-PC1	0.820	0.916	0.10
SG3-PC1	0.903	0.917	0.01
SG4-PC1	0.602	0.749	0.15
SG1-PC3	0.718	0.913	0.20
SG2-PC3	0.546	0.847	0.30

Multivariate regression models for all subseasons comparing hydrology-energy (HE) models to models incorporating logged residual data series (R_{t-1}). The asterisk identifies where R_{t-1} dominates over hydrological or energy predictors.

radiation also apparent. The change thereafter is indicated through the CPA results.

In considering the CPA divisions based on PC3, which indicates changes in the SSC characteristics, results are contrasting. At both Austre Brøggerbreen and Stagnation Glacier, no subseasonal changes are determined by the CPA. We interpret this to be an artefact of the PCA loadings. At Austre Brøggerbreen, PC3 indicates that SSC acts almost independently from all other variables considered, and examination of the graph (Fig. 3) indicates a relatively consistent score pattern. The peaks in PC3 score are short in duration, and therefore are unlikely to be interpreted as significant changes in the hydrological system. At Stagnation Glacier, the score time series is more oscillatory than at Austre Brøggerbreen, although no trends are visually obvious. PC3 at Stagnation Glacier shows SSC is linked inversely to Q, and this relation is likely to explain the fluctuations. The lack of trend indicates why no change point is identified.

The polythermal Midre Lovénbreen data set for PC3 exhibited divisions on JD186, 206, and 217. The subseasons relate to the decline in melt occurring between JD206 and 217, which effectively reduced the efficiency of the drainage. Following JD217, renewed snowmelt and raised Q was noted despite low SSC, indicating a subtle change in the drainage structure. The divide on JD186 corresponded to an increasing trend in Q, following the cooling period ending on JD184. The breakthrough event on JD198 is not picked up using this PC score series, and therefore suggests that variability in SSC between JD198 and JD206 can be explained by increasing AT and Q. Subsequently, on JD206 the variations in SSC do not relate to AT (or Q). Here, elevated SSC is observed despite lowered AT and reduced Q.

These results suggest in-stream hydrological conditions are dependent on meteorological conditions, a finding asserted by Hodson *et al.* (1998b). Further, the combination of PCA and CPA is an effective and accurate method to subdivide a time series in an objective manner.

Hydrological multivariate regression models

In examining the MRMs for Austre Brøggerbreen and considering the PC1 subseasons we infer that the initial period weakly exhibits exhaustion of sediment ($-\Sigma Q$). Sediment delivery following JD181 ($+Q^2$) is dominated by an increasing hydraulic radius, and potential thermo-erosion at the stream banks. The positive coefficients of HQ for JD200–206 and JD210–219 suggest that high flows were capable of transporting increasing quantities of sediment. Such increases have been reported by Bogen (1991), Gurnell *et al.* (1994) and Hodson & Ferguson (1999) and are interpreted to be sediment supply controlled by the thermal regime of sediments. The seasonal thaw of the active layer in the ice-proximal environment as well as the degradation of ice-cored moraine forms enables the release of sediments to the proglacial stream. Thaw occurs despite the snowfall (JD210), and this may tentatively be linked to the thermal role of meltwater as the snow both initially falls, and subsequently commences melting, or possibly the insulating properties of the snow. The brief exhaustion of sediment (JD206–210) suggests there may be some limits to sediment supply. These may be linked to stabilization of channel margins, or a decline in periglacial thaw inputs with falling temperatures. The renewed snowmelt period at the close of the season is characterized by supply from the channel margins and potential periglacial inputs.

Since no subseasons were identified for Austre Brøggerbreen based on PC3, one model was run, including all data. The predictor identified as most significant was Q^2 with a positive coefficient, showing sediment supply for the entire season at Austre Brøggerbreen was partially dependent upon channel marginal processes, with subtle changes in Q leading to large variations in SSC.

For Midre Lovénbreen, subseasons based on PC1, the early part of the season was characterized by channel margin and thaw processes $(+Q^2)$, as might be expected during periods of channel growth and stabilization. During JD200–206 it was not possible to form a 'hydrology only' model, suggesting that, as the subglacial upwelling was evolving and becoming more efficient, the sediment supply processes were truly stochastic, and not consistently related to Q in any way. For subseasons occurring after JD206, the periods are characterized by $-\Sigma Q$, implying exhaustion of the subglacial reservoir supply source. Similar exhaustion has been seen at other polythermal glaciers including Finsterwalderbreen (Hodson & Ferguson 1999). The implication of this finding is that some form of subglacial evolution from distributed to discrete drainage occurs at Midre Lovénbreen, and this glacial process dominates SSC dynamics following upwelling.

Examining models for Midre Lovénbreen, based on PC3 subseasons, the relation between SSC and Q directly $(+\log Q)$ prior to JD185 indicates that mobilization is directly linked to volume of meltwater passing through the channel. For JD186–205, SSC is linked strongly to Q directly. Prior to subglacial upwelling, this suggests instantaneous forcing dictated SSC. Following upwelling, the data may imply a link to the establishment of a subglacial channel, initially at artesian pressure, with erosion potential simply dependent on Q. After JD205, subglacial exhaustion is evident until renewed snowmelt leads to significance of HQ, suggestive of channel marginal processes dominating SSC dynamics.

Stagnation Glacier shows that for the early part of the season, SSC relates to Q. At the time following the supraglacial upwelling (JD183) sediment transport appears to exhibit a subseasonal increase $(+\Sigma Q)$. This indicates the release of sediment from an increasing source area, including both the subglacial reservoir and the enlarging drainage basin as the snowpack retreats. Increasing sediment supply from polythermal Erdmannbreen has been documented (Hodson & Ferguson 1999) for an entire

season. However, due to the intense rain event, it is unclear whether the increase seen for a subseason would have continued. Therefore, inferences regarding the subglacial system cannot be made. Following the storm on JD189, exhaustion of sediment is apparent $(-\Sigma Q)$ and latterly, $\log Q$ becomes the significant predictor for the remainder of the season. We suggest that changes in the subglacial system coincident with effective sediment supply processes in the ice-proximal environment explain these patterns. The high intensity rain storm occurring shortly after the establishment of a subglacial drainage route forced the evolution of the subglacial system. Rainfall events have been shown to lead to the genesis of a hydraulically efficient subglacial drainage system (see Warburton & Fenn 1994; Barrett & Collins 1997; Denner *et al.* 1999). We propose this occurred at Stagnation Glacier, with a shift from a distributed towards a discrete drainage system coupled to the flushing of sediment from the ice-bed interface. The lack of diurnal variability and decline in SSC immediately following the rain event (see Fig. 2) imply that such a reorganization and flushing may have occurred, and changes in Q were ineffective in mobilizing more subglacial sediments. Thereafter, sediment supply was predominantly from ice-marginal locations.

Hydrology – energy multivariate regression models

Examination of the MRMs incorporating energy terms shows an improvement in the predictive capability to a degree. While the majority of cases exhibit increases in coefficients of determination of <0.1, there is reason to suggest that thermo-erosional processes, despite not dominating the sediment delivery, may be contributory. In all cases, other than the four documented below, the inclusion of energy terms does not change the most significant predictor variable.

Four of the subseasons examined displayed a dominance of a radiation term: AB1-PC1, ML2-PC1, ML1-PC3 and SG1-PC1 (where glacier names are abbreviated, and the subseason PC is indicated). These correspond respectively to the early season, channel opening at Austre Brøggerbreen, the initial period following subglacial upwelling at Midre Lovénbreen, and upglacier enlargement of drainage basin at Midre Lovénbreen and Stagnation Glacier as the snow pack retreated to higher elevations. For AB1-PC1, thaw-related sediment delivery replaces sediment exhaustion; although apparently contradictory, these processes are not mutually exclusive. For this time, although the general tendency is for a decreasing suspended sediment load, the actual release of sediment may relate to periods of increased air temperature. This perhaps indicates channel stabilization and reduction of readily available sediments, but times of enhanced sediment load correspond to thaw related ice-marginal supply. At Stagnation Glacier, thermo-erosional and bank processes are evoked at the stream and glacier margins by the link between NR and SSC. Clearly, NR is not independent of Q, thus explaining the decline in significance of $\log Q$ as a predictor. Interpretation for Midre Lovénbreen suggests thaw- and melt-related processes may occur between JD182 and 185, as occurred at Stagnation Glacier. However, ML1-PC3 is not as clear: the inverse relation between SSC and IR for ML2-PC1 may indicate a lag between peak melt, water temperature and subglacial outflow. Subglacial erosion perhaps occurs where meltwater breaches the cold ice margin either at the ice–bed interface or by passing through sediments subject to permafrost. These interpretations are similar to those drawn from the hydrology-only MRMs.

Residual multivariate regression models

Drawing from the MRMs accounting for autocorrelation, the influence of R_{t-1} shows that SSC is linked to the flux of sediment during the previous hour. This dependency has been reported elsewhere (see Gurnell *et al.* 1994; Hodgkins 1999; Hodson & Ferguson 1999). The conclusion drawn here is that random sediment supply processes lead to slugs passing through the glaciofluvial system. However, the passage of these slugs is attenuated, with durations of over an hour. Interestingly, there is evidence to suggest periods of time are characterized by glacial, fluvial or periglacial processes (as seen with the hydrology and energy MRMs), but the dynamism of SSC during some subperiods may be more effectively explained by sediment supply processes not included in the models presented here. This suggests that, in some Arctic basins, SSC may have an autogenic tendency not accounted for by considerations of thermal and hydrological processes.

Synopsis

In comparing the success of the various models, the cumulative residual sum of squares (RSS) can be examined (see Table 3). For all glaciers MRMs based on subseasons defined by PC1 lead to the lowest RSS, and therefore have the greatest predictive potential. We suggest that, where a glacier system is effectively linked to the surrounding slope and ice-proximal environments, subseasons defined by PC1 (the seasonal trends) are more appropriate. Furthermore, we contend that the glaciofluvial output from small, high latitude glacier basins are effectively coupled to or affected by slope and periglacial processes. In these basins, subaerial processes are more significant for SSC dynamics, and therefore meteorological conditions are of direct importance. This assertion of glacial–fluvial–periglacial coupling has been made previously for Svalbard basins following analysis of proglacial terrain elevation change and glacial hydrology (see Etzelmüller 2000; Etzelmüller *et al.* 2000). Importantly, the subtle changes in processes contributing to sediment supply are closely linked to specific hydrometeorological conditions, and therefore change over time.

Note should be made of methods by which possible improvements to the MRMs presented here could be undertaken. One of the flaws in the MRMs presented here is no predictor variable is 'best matched' to SSC. Cross correlation between Q and SSC in previous studies (e.g., Gurnell *et al.* 1994; Hodson & Ferguson 1999) demonstrated the tendency for lag to exist between the two variables in Arctic basins. This lag can be attributed to hysteretic behaviour. Although as Hodgkins (1999) noted, despite hysteresis, a lag between Q and SSC may not be identified. Use of the δQ variable to account for hysteresis may be insufficient. SSC may not necessarily relate to the rate of change of discharge, but rather to the actual magnitude of discharge at a given time lag. These lags may also change over the course of season, for example as the supraglacial drainage becomes more effective with the decline in snowpack. Nevertheless, physical interpretation of lagged variables potentially becomes more problematic, particularly where directly related variables (e.g., Q and AT) may appear out of phase, or positive/negative correlations are asymmetric as a response to variability in differing data sources (for example where AT is highly variable during a partially cloudy day, but more constant at 'night').

To improve upon these subseasonal models, we suggest examining the lags between variables and incorporating these lags into MRMs may increase coefficients of determination to some extent.

Summary and implications

The temporal characteristics of SSC dynamics in three immediate Arctic proglacial streams have been compared. The combination of PCA and CPA, in sequence, is shown to be an effective and objective method to divide a multivariate time series. The CPA-derived divisions in the time series corroborated observational evidence for changes in both meteorological and hydrological conditions. The statistical technique is a tenable alternative to subjective meteorological characterization. PC1 represents seasonal trends in the data set while PC2 and PC3 may portray links between variables which explain independent modes of variance within the entire data set.

Subseasonal division was effected by the CPA using the time series of standardized scores for PC1 and PC3. The SSC dynamics was modelled for each subseason through the use of multivariate regression. Interpretation of the hydrological models provides evidence for thermo-erosional processes of sediment delivery to the fluvial channel. Incorporation of energy terms (AT, NR and IR) improves all the models presented here for predicting SSC, with substantial improvement in four cases. This improvement for all models suggests that thaw-related processes of sediment delivery may be contributory in Arctic glacial streams during different time periods, but particularly at the start of the ablation season.

Comparison of the subseasons using MRMs shows a tendency for changes between fluvial, glacial and periglacial processes delivering sediment over the course of the season. Such changes are more apparent at Austre Brøggerbreen and Stagnation Glacier, where ice marginal processes appear to be more significant. Midre Lovénbreen demonstrated the controlling dominance of the subglacial reservoir drainage and subglacial drainage evolution, which override the sediment inputs from periglacial sources. Stagnation Glacier appears to have exhibited rapid subglacial drainage evolution, forced by a rain storm, although this can not be directly confirmed with insufficient evidence available to support such an assertion. In all three basins, SSC showed true autocorrelation, with sediment sources taking periods of over 1 h to become exhausted.

Subseasons based on divisions identified through PC1 were of greater use than those from PC3 in modelling SSC dynamics. Comparison of the subseasonal model successes, given as the coefficients of determination, for MRMs based on PC1 and those derived from PC3 indicates small glaciated valleys at high latitudes may exhibit effective coupling between the slope and glaciofluvial systems. We suggest that periglacial sediment delivery processes may be significant during specific hydrometeorological conditions. Sediment supply from moraines, ice-cored terrain, thermo-erosion of permafrost and the ablation of proglacial icings are implied.

Our interpretation suggests that thermal processes do appear to govern sediment availability in Arctic glaciated basins during specific time periods (see Hodson & Ferguson 1999). Thus, there is apparent linkage between glacial, fluvial and periglacial systems. Such an interlinkage deserves closer scrutiny, particularly with projected climatic warming in high latitudes, as an increasing active layer depth may lead to enhanced subsurface transport and water flow (see Liu *et al.* 2003). However, the importance of glacial, fluvial and periglacial contributions to SSC change over the ablation season. These changes, being closely linked to hydrometeorological conditions, are likely to explain inter-annual variations in sediment yield from high latitude, glacierized basins, such as those reported by Hodgkins *et al.* (2003). Environmental change may, therefore, have significant repercussions on thermal, ice-proximal processes and associated sediment transport in Arctic glaciated basins. The periglacial influences and changing dominant processes of sediment delivery should not be ignored. In particular, these findings may have ramifications in glaciolacustrine and glaciomarine sedimentary records, and thermal process contributions should be accounted for when using of sedimentary records to reconstruct landscape denudation or palaeoclimate.

The following sources of financial and logistical support are acknowledged: NERC, Department of Indian Affairs and Northern Development (NSTP), CFI, NSERC, University of Cambridge (Worts Travelling Scholars Fund, Scandinavian Studies Fund, Department of Geography, Fitzwilliam College), Canadian Memorial Foundation, Alberta Ingenuity Fund, Polar Continental Shelf Project, Norsk Polarinstitutt, Kings Bay Kull Compani, and Parks Canada. Also, thanks to the communities of Ny Ålesund, Spitsbergen and Pond Inlet, Nunavut. Further assistance and support was provided by Brenda Mottle, Nick Cox, Adrian Hayes, Phil Hughes, Faron Anslow and Peter Wynn. Two anonymous reviewers are thanked for comments improving the clarity of this paper.

References

ASTAKHOV, V. I. 1992. The last glaciation in West Siberia. *Sveriges Geologiska Undersökning*, **81**, 21–30.

ASTAKHOV, V. I. & ISAYEVA, L. L. 1988. The 'ice hill': an example of 'retarded glaciation' in Siberia. *Quaternary Science Reviews*, **7**, 29–40.

BARRETT, A. P. & COLLINS, D. N. 1997. Interaction between water pressure in the basal drainage system and discharge from an Alpine glacier before and during a rainfall induced subglacial hydrological event. *Annals of Glaciology*, **24**, 288–292.

BJÖRNSSON, H., GJESSING, Y., HAMRAN, S.-E., HAGEN, J., LIESTØL, O., PÁLSSON, F. & ERLINGSSON, B. 1996. The thermal regime of sub-polar glaciers mapped by multi-frequency radio-echo sounding. *Journal of Glaciology*, **42**, 23–32.

BOGEN, J. 1991. Erosion and sediment transport in Svalbard. *In*: GJESSING, Y., HAGEN, J. O., HASSEL, K. A., SAND, K. & WOLD, B. (eds) *Arctic Hydrology: Present and Future Tasks*, Norwegian National Committee for Hydrology Report no. 23, 147–158.

CHEN, J. & GUPTA, A. K., 2000. *Parametric Statistical Change Point Analysis*, Birkäuser, Boston, MA, 192.

CLIFFORD, N. J., RICHARDS, K. S., BROWN, R. A. & LANE, S. N. 1995. Laboratory and field assessment of an infrared turbidity probe and its response to particle size variation in suspended sediment concentration. *Hydrological Sciences Journal*, **40**, 771–791.

DAVIS, J. 1986. *Statistics and Data Analysis in Geology*, 2nd edn., Wiley, New York, 646.

DENNER, J. C., LAWSON, D. E., LARSON, G. J., EVERSON, E. B., ALLEY, R. B., STRASSER, J. C. & KOPCZYNISKI, S. 1999. Seasonal variability in hydrologic-system response to intense rain events, Matanuska Glacier, Alaska, USA. *Annals of Glaciology*, **28**, 267–271.

ETZELMÜLLER, B. 2000. Quantification of thermo-erosion in proglacial areas – examples from Svalbard. *Zeitschrift für Geomorphologie N.F.*, **44**, 343–361.

ETZELMÜLLER, B., HAGEN, J. O., VATNE, G., ØDEGÅRD, R. S. & SOLLID, J. L. 1996. Glacier debris accumulation and sediment deformation influenced by permafrost: examples from Svalbard. *Annals of Glaciology*, **22**, 53–62.

ETZELMÜLLER, B., ØDEGÅRD, R. S., VATNE, G., MYSTERUD, R. S., TONNING, T. & SOLLID, J. L. 2000. Glacial characteristics and sediment transfer system of Longyearbreen and Larsbreen, western Spitsbergen. *Norsk Geografisk Tidsskrift*, **54**, 157–168.

EVANS, D. J. A. & ENGLAND, J. 1992. Geomorphological evidence of Holocene climatic change from northwest Ellesmere Island, Canadian high arctic. *The Holocene*, **2**, 148–158.

EVANS, M. 1997. Temporal and spatial representativeness of alpine sediment yields: Cascade Mountains, British Columbia. *Earth Surface Processes and Landforms*, **22**, 287–295.

FRENCH, H. M. & HARRY, D. G. 1990. Observations on buried glacier ice and massive segregated ice, western arctic coast, Canada. *Permafrost and Periglacial Processes*, **1**, 31–43.

GURNELL, A. M., CLARK, M. J. & HILL, C. T. 1992. Analysis and interpretation of patterns within and between hydrometeorological time series in an Alpine glacier basin. *Earth Surface Processes and Landforms*, **17**, 821–839.

GURNELL, A. M., HODSON, A. J., CLARK, M. J., BOGEN, J., HAGEN, J. O. & TRANTER, M. 1994. *Water and Sediment Discharge from Glacier Basins: and Arctic and Alpine Comparison*, IAHS Publication no. 224, 325–334.

HAAG, I. & WESTRICH, B. 2002. Processes governing river water quality identified by principal component analysis. *Hydrological Processes*, **16**, 3113–3130.

HAGEN, J. & LIESTØL, O. 1990. Long term glacier mass balance investigations in Svalbard, 1950–1988. *Annals of Glaciology*, **14**, 102–106.

HAGEN, J. O., KORSEN, O. H. & VATNE, G. 1991. Drainage pattern in a sub polar glacier: Brøggerbreen, Svalbard. *In*: GJESSING, Y., HAGEN, J. O., HASSEL, K. A., SAND, K. & WOLD, B. (eds) *Arctic Hydrology: Present and Future Tasks*, Norwegian National Committee for Hydrology Report no. 23, 121–131.

HANNAH, D. M., GURNELL, A. M. & MCGREGOR, G. R. 1999. A methodology for investigation of the seasonal evolution in proglacial hydrograph form. *Hydrological Processes*, **13**, 2603–2621.

HJELLE, A. 1993. *Geology of Svalbard*, Norsk Polarinstitutt Handbook 7.

HODGKINS, R. 1996. Seasonal trend in suspended-sediment transport from an Arctic glacier, and implications for drainage-system structure. *Annals of Glaciology*, **22**, 147–151.

HODGKINS, R. 1999. Controls on suspended-sediment transfer at a high-Arctic glacier, determined from statistical modelling. *Earth Surface Processes and Landforms*, **24**, 1–21.

HODGKINS, R., COOPER, R., WADHAM, J. & TRANTER, M. 2003. Suspended sediment fluxes in a high-arctic glacierised catchment: implications for fluvial sediment storage. *Sedimentary Geology*, **162**, 105–117.

HODSON, A. J. & FERGUSON, R. I. 1999. Fluvial suspended sediment transport from cold and warm-based glaciers in Svalbard. *Earth Surface Processes and Landforms*, **24**, 957–974.

HODSON, A. J., TRANTER, M., DOWDESWELL, J. A., GURNELL A. M., & HAGEN, J. O. 1997. Glacier thermal regime and suspended-sediment yield: a comparison of two high-Arctic glaciers. *Annals of Glaciology*, **24**: 32–37.

HODSON, A. J., GURNELL, A. M., TRANTER, M., BOGEN, J., HAGEN, J. O. & CLARK, M. J. 1998a. Suspended sediment yield and transfer processes in a small High-Arctic glacier basin, Svalbard. *Hydrological Processes*, **12**, 73–86.

HODSON, A. J., GURNELL, A. M., WASHINGTON, R., TRANTER, M., CLARK, M.J. & HAGEN J. O. 1998b. Meteorological and runoff time-series characteristics in a small high-Arctic glaciated basin, Svalbard. *Hydrological Processes*, **12**, 509–526.

HODSON, A. J., TRANTER, M. & VATNE, G. 2000. Contemporary rates of chemical denudation and atmospheric CO_2 sequestration in glacier basins:

an arctic perspective. *Earth Surface Processes and Landforms*, **25**, 1447–1471.

HODSON, A. J., TRANTER, M., GURNELL, A. M., CLARK, M. J. & HAGEN, J. O. 2002. The hydrochemistry of Bayelva, a High Arctic proglacial stream in Svalbard. *Journal of Hydrology*, **257**, 91–114.

HOLMLUND, P., BURMAN, H. & ROST, T. 1996. Sediment-mass exchange between turbid meltwater streams and proglacial deposits of Storglaciären, northern Sweden. *Annals of Galciology*, **22**, 63–67.

INLAND WATERS BRANCH. 1969. *Glacier Atlas of Canada: Bylot Island Glacier Inventory Area 46201*. Department of Energy, Mines and Resources, Ottawa.

JACKSON, G. D., DAVIDSON, A. & MORGAN, W. C. 1975. *Geology of the Pond Inlet Map Area, Baffin Island, District of Franklin*. GSC Paper 74–25, 1–33.

KAPLANSKAYA, F. A. & TARNOGRADSKIY, V. D. 1986. Remnants of the Pleistocene ice sheets in the permafrost zone as an object for palaeoglaciological research. *Polar Geography and Geology*, **10**, 257–266.

LEFAUCONNIER, B. & HAGEN, J. O. 1990. Glaciers and climate in Svalbard: statistical analysis and reconstruction of the Brøggerbreen mass balance for the last 77 years. *Annals of Glaciology*, **14**, 148–152.

LIESTØL, O. 1977. Pingos, springs and permafrost in Spitsbergen. *Norsk Polarinstitutt, Årbok*, **1975**, 7–29.

LIU, J., HAYAKAWA, N., LU, M., DONG, S. & YUAN, J. 2003. Hydrological and geocryological response of winter streamflow to climate warming in Northeast China. *Cold Regions Science and Technology*, **37**, 15–24.

LORRAIN, R. D. & DEMEUR, P. 1985. Isotopic evidence for relic Pleistocene glacier ice on Victoria Island, Canadian Arctic Archipelago. *Arctic and Alpine Research*, **17**, 89–98.

MOORMAN, B. J. & MICHEL, F. A. 2000a. The burial of ice in the proglacial environment on Bylot Island, Arctic Canada. *Permafrost and Periglacial Processes*, **11**, 161–175.

MOORMAN, B. J. & MICHEL, F. A. 2000b. Glacial hydrological system characterization using ground penetrating radar. *Hydrological Processes*, **14**, 2645–2667.

MOREHEAD, M. D., SYVITSKI, J. P. M., HUTTON, E. W. H. & PECKHAM, S. D. 2003. Modeling the inter-annual and intra-annual variability in the flux of sediment in ungauged river basins. *Global Planetary Change*, **39**, 95–110.

NOLAN, B. T. 1999. Ground water quality: nitrate behaviour in ground waters of the southeastern USA. *Journal of Environmental Quality*, **28**, 1518–1527.

PETERSEN, W., BERTINO, L., CALLIES, U. & ZORITA, E. 2001. Process identification by principal component analysis of river water-quality data. *Ecological Modelling*, **138**, 193–213.

RICHARDS, K. S. 1984. Some observations on suspended sediment dynamics in Storbregrova, Jotunheimen. *Earth Surface Processes and Landforms*, **9**, 101–112.

RICHARDS, G. & MOORE, R. D. 2003. Suspended sediment dynamics in a steep, glacier-fed mountain stream, Place Creek, Canada. *Hydrological Processes*, **17**, 1733–1753.

RIPPIN, D., WILLIS, I., ARNOLD, N., HODSON, A., MOORE, J., KOHLER, J. & BJÖRNSSON, H. 2003. Changes in geometry and subglacial drainage of Midre Lovénbreen, Svalbard, determined from digital elevation models. *Earth Surface Processes and Landforms*, **28**, 273–298.

ROBINSON, S. D., MOORMAN, B. J., JUDGE, A. S., DALLIMORE, S. R. & SHIMELD, J. W. 1992. The application of radar stratigraphic techniques to the investigation of massive ground ice at Yaya Lake, Northwest Territories. *Muscox*, **39**, 39–49.

SÁNCHEZ-MARTOS, F., JIMÉNEZ-ESPINOSA, R. & PULIDO-BOSCH, A. 2001. Mapping groundwater quality variables using PCA and geostatistics: a case study of Bajo Andarax, southeastern Spain. *Hydrological Sciences Journal*, **46**, 227–242.

SOLLID, J. L., ETZELMÜLLER, B., VATNE, G. & ØDEGÅRD, R. S. 1994. Glacial dynamics, material transfer and sedimentation of Erikbreen and Hannabreen, Liefdefjorden, northern Spitsbergen. *Zeitschrift für Geomorphologie N.F.*, **97**(suppl.), 123–144.

SYVITSKI, J. P. M. 2002. Sediment discharge variability in Arctic rivers: implications for a warmer future. *Polar Research*, **21**, 323–330.

TAYLOR, W. A. 2000. *Change-Point Analysis: a Powerful New Tool For Detecting Changes*; www.variation.com/cpa/tech/pattern.html (updated February 2000; accessed: February 2003).

VAIKMÄE, R., MICHEL, F. A. & SOLOMATIN, V. I. 1993. Morphology, stratigraphy and oxygen isotope composition of fossil glacier ice at ledyanaya gora, northwest Siberia, Russia. *Boreas*, **22**, 205–213.

VANDAELE, W. 1983. *Applied Time Series and Box-Jenkins Models*, Academic Press, New York, 417.

VATNE, G., ETZELMÜLLER, B., ØDEGÅRD, R. S. & SOLLID, J. L. 1992. Glaciofluvial sediment transfer of a sub-polar glacier, Erikbreen, Svalbard. *Stuttgarter Geographische Studien*, **117**, 253–266.

WARBURTON, J. & FENN, C. R. 1994. Unusual flood events from an Alpine glacier: observations and deductions on generating mechanisms. *Journal of Glaciology*, **40**, 176–186.

WILLIS, I. C., RICHARDS, K. S. & SHARP, M. J. 1996. Links between proglacial stream suspended sediment dynamics, glacier hydrology and glacier motion at Midtdalsbreen, Norway. *Hydrological Processes*, **10**, 629–648.

YOUNG, P. C., NG, C. N., LANE, K. & PARKER, D. 1991. Recursive forecasting, smoothing and seasonal adjustment of non-stationary environmental data. *Journal of Forecasting*, **10**, 57–89.

Melt rates at calving termini: a study at Glaciar León, Chilean Patagonia

ELEANOR HARESIGN & CHARLES R. WARREN

School of Geography and Geosciences, University of St Andrews,
St Andrews, Fife KY16 9AL, Scotland, UK
(e-mail: ech@st-andrews.ac.uk)

Abstract: Glaciar León, a temperate, grounded outlet of the North Patagonian Icefield, terminates at an active but stable calving margin in Lago Leones. Glaciological and limnological data gathered during 2001 and 2002 are used to examine the relative contributions of calving and melting to mass loss at the terminus, and the interplay between glacier and lake processes. The calving rate of 880 m/a in a mean water depth of 65 m is high for lake-calving glaciers. Subaerial melt rates at the terminus are small compared with calving rates, but melting at the waterline facilitates calving by undercutting the subaerial calving cliff. Ice-proximal surface water temperatures of 6–7°C allow waterline melt notches to grow at rates of *c.* 0.8 m/day, suggesting that melt-driven calving accounts for *c.* 23% of ice loss at the terminus. The significance of melting at calving termini decreases with increasing calving speeds, but is greater than simple calculations of melt losses suggest because of the process-linkage with calving.

The North Patagonian Icefield (46°30′–51°30′ S) is drained by about 21 large valley glaciers, around 13 of which calve into proglacial lakes and one of which calves into the sea (Aniya 2001). Most of the ice-contact lakes have formed in conjunction with glacier retreat from Little Ice Age (LIA) terminal moraines into glacially overdeepened troughs. Calving in freshwater has therefore become increasingly important to the mass balance and dynamics of the icefield (Warren & Aniya 1999). Relatively little attention has been given to the process of lacustrine calving, compared with the research effort devoted to tidewater glacier dynamics (Van der Veen 2002). Calving rates in freshwater are much slower than in comparable tidewater settings, and calving mechanisms are distinctively different (Warren *et al.* 1995a; Warren & Kirkbride 2003). At all calving glaciers, the position of the terminus is controlled by a range of interacting factors, primarily the ice flux to the terminus, the topographic geometry of the lake or fjord, calving rates, sedimentation rates and melting of the subaerial and subaqueous portions of the terminal ice cliff. Melting has usually been included implicitly in calculated calving rates, because, compared with mechanical losses by calving, melt rates have been considered to represent an insignificant contribution to mass loss at the terminus (Powell 1988; Dowdeswell &

Murray 1990). However, the validity of this generalization is undermined by evidence which suggests that melting is an important process, both at tidewater termini (Walters *et al.* 1988; Warren *et al.* 1995b; Vieli *et al.* 2002) and at lake-calving glaciers (Warren & Kirkbride 2003). The significance of melting lies not only in its direct contribution to mass losses at calving termini (Motyka *et al.* 2003), but also in its role as a trigger for mechanical calving through the creation of thermo-erosional notches at the waterline which undercut the ice cliff and cause ice slabs to break off (Kirkbride & Warren 1997; Purdie & Fitzharris 1999). Furthermore, Hanson & Hooke (2000) have pointed out that there may be sufficient energy in ice-contact water bodies to melt ice at rates in the range of observed calving speeds. The thermal regime of the ice-contact water body may therefore be an important influence on melt rates and calving processes.

Owing to the difficulties and dangers of obtaining direct measurements at calving fronts, little is understood about the interplay between the processes of melting and calving, their relative contributions to overall mass loss and their seasonal variability. Moreover, there is still no satisfactory melt rate equation that can be applied to temperate calving glaciers. A number of temperature-dependent melt relations have been developed from field, experimental and

From: HARRIS, C. & MURTON, J. B. (eds) 2005. *Cryospheric Systems: Glaciers and Permafrost.*
Geological Society, London, Special Publications, **242**, 99–109.
0305-8719/05/$15.00   © The Geological Society of London 2005.

analytical studies, but their relevance to subaqueous melting at calving termini is questionable. For example, laboratory studies such as those by Neshyba & Josberger (1980) and Russell-Head (1980) have neglected the effects of meltwater-forced convection and water depths, and have frequently used temperature ranges that exclude the low temperatures typical of ice-proximal environments. The laboratory simulations by Eijpen *et al.* (2003) measured melt rates in low-temperature freshwater and seawater, but used simple buoyant convection not forced convection. Other studies have derived melt rates from icebergs (Weeks & Campbell 1973; El-Tahan *et al.* 1987; White *et al.* 1980), but these do not accurately reflect the complex, dynamic processes operating along a subaqueous ice cliff subjected to the turbulent convective flow driven by buoyant glacial meltwater. In the most detailed field study to date, Motyka *et al.* (2003) use heat and water balances to study submarine melting at tidewater LeConte Glacier, Alaska. They conclude that, in late summer, subaqueous melting may account for at least as much ice loss at the terminus as calving, and may strongly undercut the subaerial cliff, triggering calving. Here we report the results of glaciological and limnological studies at Glaciar León and Lago Leones, and explore the interactions between glacier and lake processes.

Site description

Glaciar León (46°46′ S, 73°13′ W) is one of eight major outlet glaciers draining the eastern side of the North Patagonian Icefield (NPI; Fig. 1). It is 11.4 km long, with an ice surface area of 44.5 km² (Aniya 1988). It descends from an icefield accumulation area at a maximum elevation of 3000 m above sea level (a.s.l.), and the equilibrium line lies at 1350 m a.s.l., indicating an accumulation area ratio of 0.67 (Aniya 1988). Its three tributaries join below an icefall at *c.* 750 m a.s.l. to form a single glacier which terminates in Lago Leones at 303 m a.s.l. (Fig. 2). The entire surface of Glaciar León from the base of the icefall to the terminus, with surface gradients of 15–17°, comprises heavily crevassed ice and is largely debris-free, other than a medial/lateral moraine which divides the north and central tributaries. The steep gradient of the glacier's lower terminus, and the fact that part of the 1.5 km-wide ice cliff has lost contact with the lake, indicates that any significant further recession would involve retreat out of the lake and the cessation of calving. In common with the broad trend of historic retreat

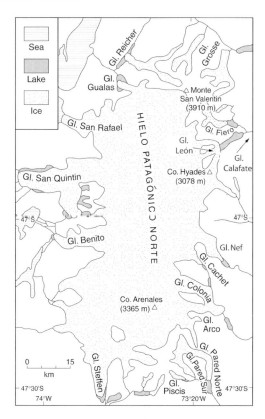

Fig. 1. Map of the North Patagonian Icefield, showing the main outlet glaciers and the location of the study area. Adapted from Glasser *et al.* (in press)

in Patagonia (Aniya & Enomoto 1986; Aniya 1999), Glaciar León retreated from a late-nineteenth-century LIA maximum in Lago Leones to its present position at the head of the lake in the mid-twentieth century. Since 1945, ice front changes have comprised small-scale fluctuations around this stable pinning point, with slow retreat prior to 1991 followed by a minor readvance (Aniya 2001). From 2001 to 2002 the average position of the terminus remained constant.

Lago Leones is 10 km long and *c.* 2.5 km wide, and has formed since the glacier margin withdrew from the 135 m-high terminal moraine that now dams the lake at its north-eastern end. Unlike most other sites around the NPI, at which ice-contact lakes formed during the late twentieth century, the onset of lake formation took place in the mid Holocene and the glacier has been calving throughout the Neoglacial (Haresign *et al.* unpublished data). Our bathymetric data show that the lake comprises several basins, defined by inferred moraine

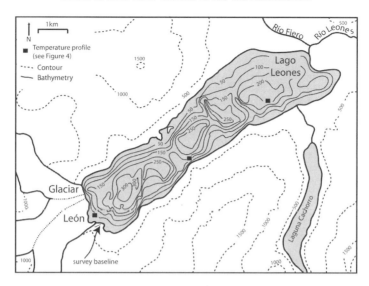

Fig. 2. Map of Lago Leones and the terminus of Glaciar León showing the location of temperature profiles in Fig. 5 and the results of the bathymetric survey (Haresign *et al.* unpublished). The baseline used for the ice velocity survey is also indicated. Map based on 1:100,000 series of the Chilean Instituto Geográfico Militar, sheets 162 and 278, and 1974 aerial photographs, updated to show the terminus position in November 2002

ridges at 3.5, 5.5 and 7 km from the glacier (Fig. 2). The maximum lake depth is about 360 m, and the width-averaged water depth at the ice front is 65 m. With an average ice cliff height of *c.* 50 m above lake level, the terminus is about 40 m above flotation.

Field methods

Fieldwork was carried out in the austral spring of 2001 and 2002. Ice surface velocities at and near the terminus were measured from 8 to 17 November 2001 and from 17 October to 11 November 2002, using a Leica TC600 total station theodolite. Because the glacier is accessible only by boat from the north-east end of Lago Leones, and because the lake can be affected by strong winds, survey frequency was weather-dependent, and the interval between surveys was roughly 2–3 days. Displacements of 33 prominent seracs across the terminus were surveyed using the triangulation method from a 69 m baseline on a rock knoll on the south (true right) side of the glacier (Fig. 2). The seracs were not evenly distributed across the glacier but divided naturally into northern, central and southern groups as a result of the interaction between glacier topography and lines of visibility from the survey baseline. The number and distribution of surveyed seracs were somewhat different in the two years because they were selected opportunistically according to the requirement that they be

unambiguously identifiable. In particular, the seracs selected in the central section of the glacier in 2002 were immediately behind the calving terminus, whereas those in 2001 were some tens of metres further back. It was not possible to account for rotation of the seracs in the survey, but this was assumed to have a negligible effect on ice velocity values because any significant rotation or shape-change of a serac rendered it unrecognizable, leading to its removal from the survey.

Survey methods were also used to quantify the development and evolution of thermo-erosional waterline notches in 2002, observed on the central part of the ice cliff where the greatest calving activity took place, supplemented by an observational record of calving events, with time intervals of *c.* 2–3 days. Ablation rates were measured over a period of 19 days in October 2002 using four stakes placed on the southern glacier margin in slow-flowing dirty ice. The contrast in debris content and flow regime means that the figures obtained provide only an upper boundary, as melt rates on the debris-free, vertical terminus cliff are likely to be lower (perhaps by 50%, cf. Kirkbride & Warren 1997).

The temperature structure of the lake was surveyed in 2001 and 2002. A total of 235 vertical temperature profiles were obtained using an electronic temperature sensor with an accuracy of $\pm 0.1^{\circ}$C and a maximum depth capability of

100 m. Temperature transects in the ice-proximal area and a long-transect of the lake were repeated several times during both field seasons to examine the development of the thermal structure of the lake during the early part of the melt season. Suspended sediment concentrations were assessed by filtering surface water samples. Measurements of precipitation, maximum and minimum temperatures and wind speed were taken using hand-held instruments, both daily at the east end of the lake, and roughly every 3 days at the glacier.

Results

Ice velocities, calving processes and calving rates

Surface ice speeds are summarized in Table 1 and vectors are plotted for each serac in Fig. 3. Ice velocities vary greatly between the three sections of the terminus, between the two years, and with distance behind the terminus, but the data yield a width-averaged velocity for the whole glacier of 880 m/a. Calving is frequent, mostly consisting of small-scale events but with occasional larger failures which produce icebergs with estimated volumes of 20,000–50,000 m^3. Subaerial calving, focused along the central portion of the ice cliff, is cyclical, progressing from thermal erosion of a melt notch at the waterline, to calving of lamellae and small ice blocks, with the cycle concluding with full-height slab calving approximately every 4 days. Waterline melting maintains a thermo-erosional notch, which undercuts the base of the subaerial cliff by as much as 6 m. Such notches are present at all times except immediately after calving events and appear to facilitate smaller calving events from the lower half of the subaerial cliff, in the build-up to a full-height slab event. Full-height calving was sometimes followed by subaqueous events which produced icebergs of similar dimensions to

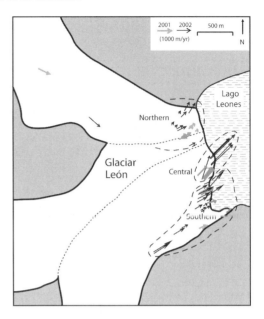

Fig. 3. Ice velocity vectors for the terminus area of Glaciar León in November 2001 and 2002. Ice front position is for November 2002

subaerially calved bergs. Smaller events consisting of disaggregated ice blocks calved mainly from the south side of the terminus. Floating ice covers very little of the lake surface beyond the immediate ice-proximal area, and all except the larger icebergs melt within 24 h. Even icebergs with long axes of tens of metres rarely reach the distal end of the lake, mostly melting within 3–4 km of the glacier. At the terminus itself, icebergs and brash ice are always present and sometimes form a continuous cover extending many tens of metres beyond the ice cliff. Although this floating ice is not static, and patches of open water within brash ice are sometimes maintained close to the ice front, there is no evidence of persistent, strong plumes of upwelling meltwater.

The calving rate (u_c in m/a) can be obtained from the equation:

$$u_c = u_i - \frac{dL}{dt} - u_m \tag{1}$$

where u_i = ice velocity, dL/dt = rate of change in glacier length (2001–2002), positive in the direction of retreat, and u_m = melt rate at the calving face, all in m/a and width-averaged. At this site, the mean terminus position is stable, i.e., $dL/dt = 0$, so the ice velocity is balanced by the sum of calving and melting, i.e., $u_i = u_c + u_m$. For comparison with published

Table 1. *Surface ice velocities of Glaciar León measured over 36 days in the austral spring of 2001 and 2002*

Glacier section	Ice velocity (m/a)		
	2001	2002	Mean
Northern	370	520	450
Central	970	1810	1390
Southern	730	870	800
Width-averaged mean	690	1070	880

'calving rates' which disregard the melt term, the value of $u_c + u_m$ at Glaciar León is therefore 880 m/a. Values of u_c for each of the three sections of the terminus were not calculated because detailed rates of terminus change of the southern and northern sections could not be accurately surveyed from the chosen baseline. However, ice velocities and terminus changes of the central section are available from 2002. In the period 17 October to 11 November 2002, the average ice velocity was 5 m/day and terminus change was 1.5 m/day, giving a calving rate ($u_c + u_m$) of 3.5 m/day. In order to derive a value for u_c alone, it is necessary to partition mass loss into its mechanically and thermally driven components. We discuss this issue below.

Limnological observations

Water temperature profiles from the ice-proximal area, mid-lake and the eastern end of Lago Leones (Fig. 2) are shown in Fig. 4. Both stratified and isothermal conditions are seen in the lake. Surface temperatures are typically 5–8°C,

cooling to about 4.5°C at 100 m depth, and there is little variation in temperature across the ice front. Lake temperature conditions evolved through October and November from near-isothermal conditions with a warm surface layer, to a thermal structure with more gradual cooling from the surface and weak thermocline development at 30–40 m deep (Fig. 5), which develops westwards towards the glacier as the season advances. Average suspended sediment values are low, at 37 mg/l, and show little spatial variation.

Meteorological observations

Results of the meteorological observations are given in Table 2. The dominant winds were katabatic winds draining down the neighbouring Fiero valley (north-west of Lago Leones), typically in the afternoon. These predominated over katabatics from Glaciar León, typically leaving

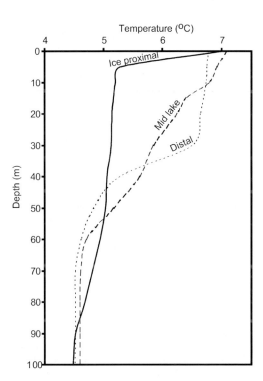

Fig. 4. Representative temperature profiles for Lago Leones, showing near-isothermal conditions in the ice-proximal waters and the development of a thermocline in the distal part of the lake

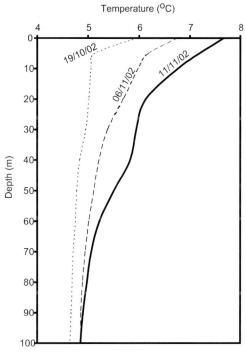

Fig. 5. Temperature profiles for Lago Leones, showing the temperature evolution of the lake during early summer. Each plot represents the average of the long profile of the lake, made up of 14 repeated vertical profiles. The lake structure progresses from a predominantly isothermal regime to one showing warming to greater depths with the presence of a thermocline at 30–40 m deep

Table 2. *Weather data from Lago Leones in 2001 (13 October to 16 November) and 2002 (15 October to 14 November), showing daily average values*

Year	Temperature (°C)		Precipitation (mm)		Number of rain days (%)
	Maximum	Minimum	Lake end	Glacier	
2001	15.5	3.5	3.0	6.5	43
2002	14.0	6.0	4.5	7.0	62

the ice-proximal area calm and free of waves. Winds of 16–22 km/h were experienced most days, though gusts of up to 45 km/h were recorded.

Discussion

Ice velocities

Surface ice velocities are known for only a few of Patagonia's lake-calving glaciers (Warren & Aniya 1999). The annual values extrapolated from our short-term surveys ignore seasonal variations but this is negligible in Patagonia (c. 5% summer–winter difference) due to high, year-round precipitation and relatively mild winters, permitting continuous basal sliding (Rott et al. 1998; Skvarca et al. 2003). Strain heating and sliding friction associated with high flow speeds will also contribute meltwater to the basal hydrological system, creating positive feedback for maintaining fast flow. The annual velocities extrapolated from our spring survey, and the derived calving rate, may therefore approximate to true annual values.

Surface ice speeds were significantly and consistently faster in 2002 than in 2001 (Table 1). The most straightforward explanation for this is the greater frequency of rain days in 2002, combined with the higher minimum temperatures (Table 2), both of which would increase meltwater volumes and enhance basal sliding. However, the apparent dramatic increase in the mean velocity (690 m/a in 2001 to 1070 m/a in 2002) is exaggerated by the doubling of speed in the central section. In turn, part of this increase may be a sampling artefact, reflecting the fact that the seracs surveyed in 2002 were closer to the terminus than those in 2001, placing them in a faster location within the zone of strong acceleration behind the calving cliff. Given the high spatio-temporal variability of the velocity field at Glaciar León, the derived mean velocity figure of 880 m/a must be regarded as provisional until more complete data are available.

Calving rates and processes

One well established statistical correlation for calving glaciers is a linear relationship between calving speed (u_c) and water depth (h_w). For the relatively small number of calving glaciers worldwide for which u_c and h_w are available, this correlation is robust, although not fully understood. However, the slope coefficient of this linear relationship varies significantly for different populations of calving glaciers: temperate, polar, freshwater, tidewater and combinations of these. In particular, u_c at temperate tidewater glaciers is typically an order of magnitude faster in any given water depth than at temperate lake-calving glaciers. For a width-averaged water depth of 65 m, as at the terminus of Glaciar León, the tidewater relationship of Brown et al. (1982) predicts $u_c = 1762$ m/a, while the freshwater relationship of Warren & Kirkbride (2003) predicts $u_c = 167$ m/a. The observed calving rate of 880 m/a at Glaciar León is therefore something of an anomaly, plotting in an intermediate position. However, given that all three tributaries have large icefalls just upstream of the current location of the terminus and that the lower glacier is steep and pervasively crevassed, the high flow speeds and high calving rate are perhaps unsurprising. Alternatively, this high value of u_c could prove not to be anomalous if further work reveals that the u_c/h_w correlation for freshwater is nonlinear. This possibility was tentatively raised by Warren et al. (1995a) but adequate data still do not exist to test it.

The cyclic pattern of subaerial calving recorded at Glaciar León, together with the associated repeating sequence of ice front geometries, closely reflect observations at lake-calving glaciers in New Zealand (Kirkbride & Warren 1997; Purdie & Fitzharris 1999). The cycle, however, is shorter, taking c. 4 days at Glaciar León compared with c. 15 days at the slower-flowing New Zealand glaciers, and is accompanied by faster change of terminus geometry. Until recently it was thought that subaerial calving driven by waterline melting was restricted to lake-calving settings. However, Vieli *et al.* (2002) report that such calving behaviour predominates at the tidewater Hansbreen, Svalbard, and that seasonal patterns of melting and calving are closely related. This suggests either that this style of subaerial calving is common to both freshwater glaciers and polar tidewater glaciers, and/or that it is a process which becomes progressively less important with increasing calving speeds. For example, subaerial calving due to waterline melting is not characteristic at the 12 tidewater glaciers in the Brown *et al.* (1982) dataset in which the mean value of u_c is 1885 m/a, whereas at the 21 lake-calving glaciers in the Warren & Kirkbride (2003) dataset the mean value of u_c is 136 m/a and melt-induced calving is common. At Hansbreen $u_c = 164$ m/a. Again, further work is necessary to resolve this issue.

Thermal structure of Lago Leones

The u_c/h_w relation has proved robust at sites around the world as a useful first-order predictor of calving speeds, but it is also apparent that the water temperature of proglacial water bodies may significantly influence calving rates and processes (Kennett *et al.* 1997; Warren 1999). For 10 lake-calving glaciers at which ice-proximal water temperature (t_w) data exist, a positive correlation exists between u_c and t_w, and the strength of this correlation is similar to that between u_c and h_w in freshwater (Warren & Kirkbride 2003). Calving dynamics at lake-calving glaciers are therefore intimately linked with the thermal regime of the lakes in which they terminate (Roehl 2003).

Available data about physical limnology in ice-contact lakes and its influence on calving termini have been reviewed and discussed by Warren & Kirkbride (1998) and Benn & Evans (1998, p. 282ff). Because few data exist, present understanding rests on a small number of studies mostly focused on sediment dynamics (e.g., Gustavson 1975; Harris 1976). Very little attention has been given to the interaction between limnological and glaciological processes, the study of Tasman Lake and Tasman Glacier by Roehl (2003) being a rare exception. Water temperatures recorded in Lago Leones are higher than in many ice-contact lakes, even though our work was carried out in spring, whereas most data in other locations have been obtained in summer or autumn. The greater warmth probably reflects the lake's relatively low latitude (46° S) and altitude (303 m), and also its large size (*c.* 20 km^2), which dilutes the influence of glacial meltwater input. Another factor is the absence of floating ice from most parts of the lake. This contrasts with many other ice-contact lakes in Patagonia where abundant icebergs prevent the development of stratification and deliver meltwater to all parts of the lake (Warren 1999; Warren *et al.* 2001).

Water in the ice-proximal area is isothermal below a depth of 5 m, presumably due to turbulent mixing driven by meltwater inputs. Evidence for meltwater input is indicated by a lake level rise, in both 2001 and 2002, of 0.8 and 0.3 m, respectively, indicating a meltwater input of $6-16 \times 10^6 \, \text{m}^3$, which returned to its original level after about 5 days. However, the existence of a warm layer in the upper 5 m and the absence of obvious zones of upwelling suggest that this mixing is not vigorous, and that the meltwater does not form a buoyant overflow plume like that seen at the tidewater LeConte Glacier, Alaska (Motyka *et al.* 2003). Ice-proximal convective cells driven by meltwater inputs from the glacier are less powerful in freshwater than in tidewater because meltwater at 0°C is 200 times more buoyant in saltwater than in freshwater, a contrast which is probably part of the explanation for faster calving rates in tidewater (Funk & Röthlisberger 1989). The temperature stratification which evolved during the late spring, with all temperatures above 3.98°C (the density extremum for sediment-free water), and temperatures warming towards the surface, is indicative of a stable water column. The lake represents a large reservoir of heat which has the potential to melt glacier ice. The question, however, is whether this thermal potential can be realised through advection of warm waters to the calving front by, for example, convection, turbulent mixing and/or wind.

Lake-glacier interactions: how important is melting?

Melting has long been presumed unimportant at calving termini, both in absolute and process terms, but recent evidence has questioned this assumption, suggesting that it is inappropriate to incorporate melting within 'calving rates'. These two components of mass loss are strongly linked at lake-calving termini where waterline melting may be the rate-controlling process for subaerial calving (Kirkbride & Warren 1997; Roehl 2003; Warren & Kirkbride 2003). It is therefore hard to separate them precisely. Nevertheless, a more complete understanding of processes operating at calving termini requires that mass loss should be partitioned into its thermal and mechanical components.

One possible way of achieving this is to employ an ice flux balance, using the difference between ice flux into the terminus and iceberg discharge from the terminus, but the uncertainties of calculating the latter are very large (Warren et al. 1995b; Motyka et al. 2003). Another, recognizing that melt rates will vary significantly above, below and at the waterline, is to consider melting in each zone of the terminus. In general terms, subaqueous melt rates are likely to be higher than subaerial melt rates, with the highest rates of all taking place at the waterline (El-Tahan et al. 1987). Each of these zones is now considered in turn at Glaciar León.

Subaerial melt rates

Estimates of the upper boundary for melt rates on the cliff face, from the ablation stake data at the glacier margin, yield a mean rate of 0.04 m/day. Whilst rates were consistent between the four stakes during each measurement period, the stakes melted out of the ice over periods longer than 24 h, limiting the ablation record to the periods when access was possible every 24 h, shown in Table 3. This average spring rate is lower than summer rates reported at New Zealand lake-calving glaciers [0.07 m/day (Kirkbride & Warren 1997); 0.08 m/day (Purdie & Fitzharris 1999); 0.06–0.1 m/day (Roehl 2003)] and at the tidewater LeConte Glacier [0.1 m/day (Motyka et al. 2003)]. A rate of 0.04 m/day implies that u_m accounts for horizontal cliff retreat of 15 m/a or just 2% of total ice loss ($u_c + u_m = 880$) from the subaerial cliff.

Waterline melting

The rate of waterline melting was assessed by surveying the size of waterline notches in October–November 2002. On the central part of the ice cliff, where the greatest calving activity took place, notches with a horizontal depth of up to 6.3 m developed. The waterline melt notches

Table 3. *Ablation rates at Glaciar León, 2002, measured at the southern margin, (23 October to 10 November), showing values obtained over 24 h periods*

Date	Ablation rate (m/day)
23–24 October	0.06
24–25 October	0.05
25–26 October	0.04
26–27 October	0.01
10–11 November	0.04

observed at Glaciar León are slightly larger than those reported at several lake-calving glaciers in New Zealand by Warren & Kirkbride (2003) but Purdie & Fitzharris (1999) report undercuts up to 10 m deep at Tasman Glacier. Unfortunately, notch growth rate could not often be accurately calculated because measured notches did not persist from one survey to the next. Consequently, although melt notch size could be measured, the rate of notch growth could be quantified for only one period, yielding a value of 0.8 m/day. This was calculated by surveying the increasing distance between the front and back of the notch over a 2-day period, during which no calving occurred. Despite the uncertainty, waterline melt rates appear to be an order of magnitude greater than inferred rates of melting on the subaerial ice cliff. These waterline rates also provide an upper limit for subaqueous rates because maximum rates of aqueous melt are known to occur at the waterline (Eijpen et al. 2003; Warren & Kirkbride 1998).

Using the calving rate ($u_c + u_m = 3.5$ m/day) and melt rate derived above for the central section of the ice cliff, where notch development is most rapid, waterline melting-driven calving would account for c. 23% of mass loss at the terminus in the spring of 2002. Lago Leones does not freeze over in winter, so waterline melting must continue throughout the year. However, since u_m will increase through the summer into the autumn, when maximum water temperatures are observed (Motyka et al. 2003), the contribution of waterline melting to the overall calving rate is likely to increase between November and February. Given that waterline melting probably accounts for less than half of mass loss at the terminus, it cannot be the primary rate-controlling process for calving at Glaciar León as it is at the slow-moving New Zealand glaciers studied by Warren & Kirkbride (2003). It nevertheless accounts for a significant fraction of subaerial calving and, by partially controlling the geometry of the calving cliff, modifies the stress distribution at the glacier terminus.

Subaqueous melt rates

These, by their very nature, are the hardest to quantify. Melt rates of glacier ice in water have been studied in a variety of non-calving settings, but the rates and variability of subaqueous melt rates at calving termini are still highly uncertain. In an enclosed lake setting, one way to consider energy transfer to the ice face is to define a heat budget for the lake, an approach used by Sakai et al. (2000). However, because this method is hampered by the difficulty of accurately accounting for all the inputs and outputs, a more direct way of calculating melt rates is required.

One approach is to derive temperature-dependent melt relations, either from laboratory studies or from observations of rates of iceberg deterioration in the open sea. Three of the best known such relations are these:

$$u_m = 2.78T + 0.47T^2 \quad (m/a)$$

$$\text{(Neshyba \& Josberger 1980)} \quad (2)$$

$$u_m = 1.8 \times 10^{-2}(T + 1.8)^{1.5} \quad (m/day)$$

$$\text{(Russell-Head 1980)} \quad (3)$$

$$u_m = 6.74 \times 10^{-6} v^{0.8} T/l^{0.2} \quad (m/s)$$

$$\text{(Weeks \& Campbell 1973)} \quad (4)$$

where T is temperature (°C), v is water current velocity [=0.1 and 0.2 m/s (Roehl 2003)], and l is iceberg length. Here, l (=65 m) is ice thickness below water, i.e., parallel to the expected dominant direction of buoyant upwelling. Using ice-proximal water temperatures from Lago Leones, equation (2) yields significantly lower values than the closely similar values from equations (3) and (4) (Table 4). Calving termini clearly represent very different contexts from either tabular icebergs melting in the open ocean or laboratory experiments on melt rates, and so there are obvious problems in attempting to apply relationships developed for the latter to calving termini. Nevertheless, in the absence of alternatives these equations have been employed to calculate subaqueous melt rates at tidewater glaciers (Powell 1983; Powell & Molnia 1989; Syvitski 1989). For example, Hunter et al. (1996) use equation (4) to calculate melt rates of 21–31 m/a for various temperate tidewater glaciers in Alaska. However, it is not always clear whether l should refer to ice thickness below water or to glacier width. Moreover, small differences in upwelling rates greatly affect the calculated value of u_m (Dowdeswell & Murray 1990), yet upwelling rates in the

boundary zone at calving fronts are poorly known and highly variable. Further doubt about the usefulness of such melt relations is raised by Motyka et al. (2003), also working at an Alaskan tidewater glacier. Their energy balance approach yields a melt rate of 12 m/day (equivalent to 57% of the total mass loss at the terminus), a rate which is greater by a factor of 175 than the rates calculated for comparable settings by Hunter et al. (1996). Such a gulf of difference between recently published melt rates is not only remarkable but emphasizes how poorly understood these processes are, even at tidewater glaciers where work has been focused. Melt rates in freshwater are even less well understood.

It seems, then, that since direct measurement is difficult and current speeds are unknown at Glaciar León, available data do not enable rates of subaqueous melting to be calculated accurately. It is probable, though, that rates of mechanical loss (through subaqueous calving) exceed melt rates, as they do along the subaerial cliff. If even waterline melting operates at c. 23% of the rate of subaerial calving, subaqueous melting can only account for a small percentage of mass loss from the submerged portion of the face at this site. Moreover, melt rates decrease with increasing water pressure at depth (Huppert 1980). Rapid calving of the subaerial cliff, leading to a projecting underwater 'ice foot', will expose submerged ice to increasing buoyant forces which will eventually trigger calving when buoyancy exceeds a critical stress for failure (cf. Hunter & Powell 1998). The observation that subaqueous icebergs are not consistently larger than subaerial ones is unusual, contrasting with the pattern typically observed at grounded calving termini, both in tidewater and freshwater (Warren et al. 1995b; Warren & Kirkbride 2003). It suggests that the basal part of the glacier is pervasively fractured, presumably through bottom crevassing driven by high flow speeds through the icefalls, so that the critical failure stress is relatively low.

This discussion makes clear how difficult it is to quantify rates of melt at calving termini and to partition total mass loss $(u_c + u_m)$ into its two

Table 4. *Ice melt rates from published equations (see text) using mean ice-proximal water temperatures measured in Lago Leones in 2001 and 2002*

Depth (m)	Temperature (°C)	Equation (2), (m/a)	Equation (3), (m/a)	Equation (4), $v = 0.1$ (m/a)	Equation (4), $v = 0.2$ (m/a)
0	6.44	37	155	94	164
5	5.91	33	141	86	150
10	5.71	31	135	84	145

primary components. At Glaciar León this means that it is not possible to derive an accurate value for u_c alone. All that can be said with confidence is that calving accounts for a majority of mass loss and melting for a small but significant minority. From the few existing lake-calving studies there is evidence that melting is of relatively greater significance (both as a process of mass loss and as a triggering mechanism for calving) when u_c is low and becomes progressively less significant as u_c increases. Thus at slow-flowing glaciers (such as those in New Zealand) waterline melting can be the rate-controlling process for calving and cliff melting may account for much of the ice loss. For example, Purdie & Fitzharris (1999) suggest that melting accounts for 71% of subaerial ice loss at the terminus of Tasman Glacier. By contrast, at sites with high flow speeds, pervasive crevassing and rapid calving, including Patagonian glaciers like Glaciar León and Glaciar Ameghino (Warren 1999), the rate of waterline notch development is slow compared with u_c. In such contexts, melt-induced calving and cliff melting only account for a small portion of mass loss at the terminus. The end member of this spectrum may be represented by deep-water sites such as Glaciar Upsala, Argentina (Skvarca et al. 2002, 2003) or Glaciar Nef, Chile (Warren et al. 2001), where calving driven by buoyant forces at floating or near-floating termini is of such high magnitude and is so dominant a process that it renders melting insignificant.

Conclusion

Glaciar León calves at a mean rate of 880 m/a in a mean water depth of 65 m into the relatively warm (4.5–8°C) waters of Lago Leones. Melting along the terminal ice cliff accounts for a minority of total mass loss at the terminus. Nevertheless, by undercutting the subaerial part of the calving cliff, melting at the waterline plays a significant role in facilitating subaerial calving. This work emphasizes that the calving dynamics of lake-calving glaciers cannot be understood without detailed knowledge of the physical limnology of their ice-contact lakes. Existing data suggest that the relative significance of melting decreases with increasing calving speeds. At slow-flowing lake-calving glaciers, waterline melting may be the rate-controlling process for calving, but with increasing calving rates the proportion of mass loss that is caused by melt-induced calving declines. Subaqueous melting may also be important, but with present knowledge it is not possible to estimate its magnitude, variability or relative significance. Existing temperature-dependent melt relations, which largely relate to saltwater

environments and which ignore forced convection, are inadequate tools for calculating subaqueous melt rates, especially at freshwater sites. To improve our understanding of calving, both in lakes and in tidewater, it will be necessary to investigate the immediate ice-proximal aqueous environment in greater detail, paying particular attention to water temperatures, upwelling rates and their spatio-temporal variability.

This research was carried out with the permission of CONAF (Corporación Nacional Forestal) in Coyhaique. Many thanks go to staff and Venturers of Raleigh International, without whose logistical support and enthusiasm this work would not have been possible. EH was in receipt of funding from the University of St Andrews, the American Alpine Club, the British Geomorphological Research Group, the Dudley Stamp Memorial Trust, the Royal Scottish Geographical Society and the Quaternary Research Association. C.R.W. was supported by the Carnegie Trust for the Universities of Scotland. The authors gratefully acknowledge the helpful suggestions of S. Harrison and the detailed reviewing of B. Rea, which substantially improved the paper.

References

ANIYA, M. 1988. Glacier inventory for the Northern Patagonian Icefield, Chile, and variations 1944/45 to 1985/86. Arctic and Alpine Research, 20, 179–187.
ANIYA, M. 1999. Recent glacier variations of the Hielo Patagónicos, South America and their contribution to sea level change. Arctic, Antarctic and Alpine Research, 31, 144–152.
ANIYA, M. 2001. Glacier variation of Hielo Patagónico Norte, Chilean Patagonia, since 1944/45, with special reference to variations between 1995/1996 and 1999/2000. Bulletin of Glaciological Research, 18, 55–63.
ANIYA, M. & ENOMOTO, H. 1986. Glacier variations and their causes in the Northern Patagonian Icefield, Chile, since 1944. Arctic and Alpine Research, 18, 307–316.
BENN, D. I. & EVANS, D. J. A. 1998. Glaciers and Glaciation, Arnold, London.
BROWN, C. S., MEIER, M. F. & POST, A. 1982. Calving speed of Alaska tidewater glaciers with applications to the Columbia Glacier, Alaska, US Geological Survey Professional Paper 1258-C, 13pp.
DOWDESWELL, J. A. & MURRAY, T. 1990. Modelling rates of sedimentation from icebergs. In: DOWDESWELL, J. A. & SCOURSE, J. D. (eds) Glacimarine Environments: Processes and Sediments. Geological Society Special Publication no. 53, 121–137.
EIJPEN, K. J., WARREN, C. R. & BENN, D. I. 2003. Subaqueous melt rates at calving termini: a laboratory approach. Annals of Glaciology, 36, 179–183.
EL-TAHAN, M. S., VENKATESH, M. S. & EL-TAHAN, H. 1987. Validation and quantitative assessment of the deterioration mechanisms of arctic icebergs.

Journal of Offshore Mech. Arctic Engineering, **109**, 102–108.

FUNK, M. & RÖTHLISBERGER, H. 1989. Forecasting the effects of a planned reservoir which will partially flood the tongue of Unteraargletscher in Switzerland. *Annals of Glaciology*, **13**, 76–81.

GLASSER, N. F., HARRISON, S., WINCHESTER, V. & ANIYA, M. 2004. Late Pleistocene and Holocene palaeoclimate and glacier fluctuations in Patagonia. *Global and Planetary Change*, **43**, 79–101.

GUSTAVSON, T. C. 1975. Bathymetry and sediment distribution in proglacial Malaspina Lake, Alaska. *Journal of Sedimentary Petrology*, **45**, 450–461.

HARRIS, P. W. V. 1976. The seasonal temperature-salinity structure of a glacial lake: Jökulsárlón, south-east Iceland. *Geografiska Annaler*, **58A**, 329–336.

HANSON, B. & Hooke, LeB. R. 2000. Glacier calving: a numerical model of forces in the calving speed/water-depth relation. *Journal of Glaciology*, **153**, 188–196.

HUNTER, L. E. & POWELL, R. D. 1998. Ice foot development at temperate tidewater margins in Alaska. *Geophysical Research Letters*, **25**, 1923–1926.

HUNTER, L. E., POWELL, R. D. & LAWSON, D. E. 1996. Flux of debris transported by ice at three Alaskan tidewater glaciers. *Journal of Glaciology*, **42**(140), 123–135.

HUPPERT, H. E. 1980. The physical processes involved in the melting of icebergs. *Annals of Glaciology*, **1**, 97–101.

KENNETT, M., LAUMANN, T. & KJØLLMOEN, B. 1997. Predicted response of the calving glacier Svartisheibreen, Norway, and outbursts from it, to future changes in climate and lake level. *Annals of Glaciology*, **24**, 16–20.

KIRKBRIDE, M. P. & WARREN, C. R. 1997. Calving processes at a grounded ice cliff. *Annals of Glaciology*, **24**, 116–121.

MOTYKA, R. J., HUNTER, L., ECHELMEYER, K. & CONNOR, C. 2003. Submarine melting at the terminus of a temperate tidewater glacier, LeConte Glacier, Alaska, U.S.A. *Annals of Glaciology*, **36**, 57–65.

NESHYBA, S. & JOSBERGER, E. G. 1980. On the estimation of Antarctic iceberg melt rate. *Journal of Physical Oceanography*, **10**, 1681–1685.

POWELL, R. D. 1983. Glacial-marine sedimentation processes and lithofacies of temperate tidwater glaciers, Glacier Bay, Alaska. *In*: MOLNIA, B. F. (ed.) *Glacial-marine Sedimentation*, Plenum Press, New York, 185–232.

POWELL, R. D. 1988. *Processes and Facies of Temperate and Sub-polar Glaciers with Tidewater Fronts*, Geological Society of America Short Course Notes, Geological Society of America, Boulder, CO.

POWELL, R. D. & MOLNIA, B. F. 1989. Glacimarine sedimentary processes, facies, and morphology of the south-southwest Alaska Shelf and fjords. *Marine Geology*, **85**, 359–390.

PURDIE, J. & FITZHARRIS, B. 1999. Processes and rates of ice loss at the terminus of Tasman Glacier, New Zealand. *Global and Planetary Change*, **22**, 79–91.

ROEHL, K. 2003. Thermal regime of an ice-contact lake and its implication for glacier retreat. *In*: *Ice in the Environment: Proceedings of the 16th IAHR International Symposium on Ice*, Dunedin, New Zealand, 2–6 December 2002, 304–312.

ROTT, H., STUEFER, M., SIEGEL, A., SKVARCA, P. & ECKSTALLER, A. 1998. Mass fluxes and dynamics of Moreno Glacier, Southern Patagonia Icefield. *Geophysical Research Letters*, **25**, 1407–1410.

RUSSELL-HEAD, D. S. 1980. The melting of free-drifting icebergs. *Annals of Glaciology*, **1**, 119–122.

SAKAI, A., CHIKITA, K. & YAMADA, T. 2000. Expansion of a moraine-dammed glacier lake, Tsho Rolpa, in Rolwaling Himal, Nepal Himalaya. *Limnology and Oceanography*, **45**, 1401–1408.

SKVARCA, P., De ANGELIS, H., NARUSE, R., WARREN, C. R. & ANIYA, M. 2002. Calving rates in freshwater: new data from southern Patagonia. *Annals of Glaciology*, **34**, 379–384.

SKVARCA, P., RAUP, B. & DE ANGELIS, H. 2003. Recent behaviour of Glaciar Upsala, a fast-flowing calving glacier in Lago Argentino, southern Patagonia. *Annals of Glaciology*, **36**, 184–188.

SYVITSKI, J. P. M. 1989. On the deposition of sediment within glacier-influenced fjords: Oceanographic controls. *Marine Geology*, **85**, 301–329.

VAN DER VEEN, C. J. 2002. Calving glaciers. *Progress in Physical Geography*, **26**, 96–122.

VIELI, A., JANIA, J. & KOLONDRA, L. 2002. The retreat of a tidewater glacier: observations and model calculations on Hansbreen, Spitsbergen. *Journal of Glaciology*, **48**, 592–600.

WALTERS, R. A., JOSBERGER, E. G. & DRIEDGER, C. L. 1988. Columbia Bay, Alaska: an 'upside down' estuary. *Estuarine Coastal Shelf Science*, **26**, 607–617.

WARREN, C. R. 1999. Calving speed in freshwater at Glaciar Ameghino, Patagonia. *Zeitschrift für Gletscherkunde und Glazialgeologie*, **35**, 21–34.

WARREN, C. R. & ANIYA, M. 1999. The calving glaciers of southern South America. *Global and Planetary Change*, **22**, 59–77.

WARREN, C. R. & KIRKBRIDE, M. P. 1998. Temperature and bathymetry of ice-contact lakes in Mount Cook National Park, New Zealand. *New Zealand Journal of Geology and Geophysics*, **41**, 133–143.

WARREN, C. R. & KIRKBRIDE, M. P. 2003. Calving speed and climatic sensitivity of New Zealand lake-calving glaciers. *Annals of Glaciology*, **36**, 173–178.

WARREN, C. R., GREENE, D. & GLASSER, N. F. 1995a. Glaciar Upsala, Patagonia: Rapid calving retreat in fresh water. *Annals of Glaciology*, **21**, 311–316.

WARREN, C. R., GLASSER, N. F., HARRISON, S., WINCHESTER, V., KERR, A. R. & Rivera A. 1995b. Characteristics of tidewater calving at Glaciar San Rafael, Chile. *Journal of Glaciology*, **41**, 273–289.

WARREN, C. R., BENN, D. I., WINCHESTER, V. & HARRISON, S. 2001. Buoyancy-driven lacustrine calving, Glaciar Nef, Chilean Patagonia. *Journal of Glaciology*, **47**, 135–146.

WEEKS, W. F. & CAMPBELL, W. J. 1973. Icebergs as a freshwater source: an appraisal. *Journal of Glaciology*, **12**, 207–232.

Actual paraglacial progradation of the coastal zone in the Kongsfjorden area, western Spitsbergen (Svalbard)

DENIS MERCIER[1] & DOMINIQUE LAFFLY[2]

[1]*Department of Geography University of Paris 4 Sorbonne,
191 rue Saint Jacques 75 005 Paris, France
(e-mail: denis.mercier@paris4.sorbonne.fr)*
[2]*Department of Geography University of Pau avenue du
Doyen Poplawski 64 000 Pau, France
(e-mail: dominique.laffly@univ-pau.fr)*

Abstract: This research was carried out on the Brøgger Peninsula, northwest Spitsbergen, Svalbard (79°N 12°E). In the western part of Spitsbergen, cold-based valley glaciers have retreated more than 1 km from their Little Ice Age limits, and glacial meltwater has extensively reworked glacigenic sediments on exposed glacier forelands. In such areas, a paraglacial sediment transport regime has become predominant, with runoff as the dominant process. A combination of GIS, DEM, aerial photographic and field data was employed to estimate shoreline progradation at sandur outflows. Average shoreline progradation is estimated to amount to 3 m/annum over the last 30 years, a period of uninterrupted sediment provision from the glacial runoff system.

Since the end of the Little Ice Age, high-latitude glacier margins have undergone net retreat in response to changing meteorological conditions, with associated geomorphological consequences (Hansen 1999; Lefauconnier *et al.* 1999; Førland & Hanssen-Bauer 2000; Hanssen-Bauer & Førland 1998; Six *et al.* 2001; Hanssen-Bauer 2002; Humlum 2002; Rippin *et al.* 2003; Hagen *et al.* 2003). As a result, parts of Spitsbergen are experiencing a transition from a landscape dominated by glacial and periglacial processes to one in which paraglacial landscape response is predominant (Mercier 2000, 2002), as are other glacier margins all over the world (Ballantyne 2002). In western Spitsbergen (Kongsfjorden, Isfjorden, Bellsund, Hornsund areas) cold-based valley glaciers have retreated more than 1 km from their Little Ice Age maxima (Hagen *et al.* 1993), and glacigenic sediments exposed by glacier retreat are being extensively reworked by proglacial meltwater streams. This paper employs GIS-based analysis of geomorphological changes that have occurred since the end of the Little Ice Age to demonstrate how seaward progradation of the outwash plain is associated with periods of high glacifluvial sediment supply to the shoreface, whereas coastal erosion and shoreline recession predominate when glacifluvial sediment flux is reduced.

Study area

Field investigations were carried out on the northern part of the Brøgger Peninsula, NW Spitsbergen (78°55′ N, 12°10′ E, Fig. 1). The study area is mountainous and supports a number of valley glaciers, like midre Lovénbreen, austre Lovénbreen and vestre Lovénbreen. Active sandurs (outwash plains) occupy the zone between arcuate terminal moraine complexes and fjord coastline (Fig. 2).

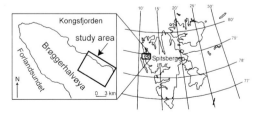

Fig. 1. Map showing Svalbard and the location of the study area in West Spitsbergen

From: HARRIS, C. & MURTON, J. B. (eds) 2005. *Cryospheric Systems: Glaciers and Permafrost.*
Geological Society, London, Special Publications, **242**, 111–117.
0305-8719/05/$15.00 © The Geological Society of London 2005.

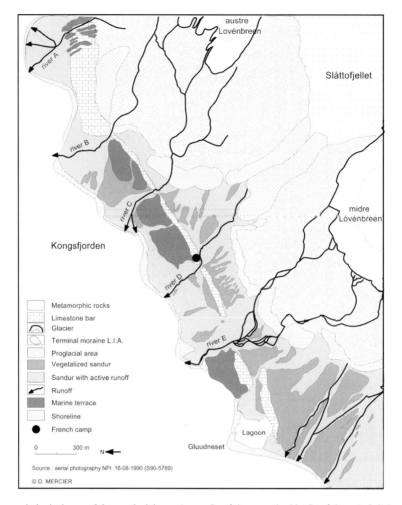

Fig. 2. Geomorphological map of the proglacial area (austre Lovénbreen and midre Lovénbreen), Spitsbergen

The barrier coastline at the seaward end of the outwash plains constitutes a paraglacial coast in which the nearshore sediment budget is dominated by inputs of reworked glacigenic sediments transported by glacial meltwater streams. Forbes & Syvitski (1994) defined paraglacial coasts 'to be those on or adjacent to formerly ice-covered terrain, where glacially excavated landforms or glacigenic sediments have a recognizable influence on the character and evolution of the coast and nearshore deposits'.

Mass-balance investigations on the local glaciers indicate a negative net balance since the end of the Little Ice Age (Lefauconnier *et al.* 1999; Nesje & Dahl 2000). Small glaciers (around 8 km² in area) have been retreating throughout the twentieth century, although in 1907 austre Lovénbreen, midre Lovénbreen and

vestre Lovénbreen were at their maximum, with domed fronts associated with the outermost frontal moraines, as we can see on Isachsen's photos (Isachsen 1912). In 1936, these glaciers were still very close to their terminal moraines (Fig. 3). Rippin *et al.* (2003) calculated mean annual mass balance of −0.61 m/a (water equivalent) for Midre Lovénbreen between 1977 and 1985. This glacier retreat is correlated with the current climatic change. During 1975–1996 in Ny-Ålesund, 25% of the precipitation was rain, 44% was snow and 31% was mixed, sleet or combination of rain and snow (Førland & Hanssen-Bauer, 2000), with the trend for snow being negative and that for mixed precipitation positive. The annual mean temperature at Svalbard Airport has increased by 0.14°C per decade since 1912 and annual precipitation has

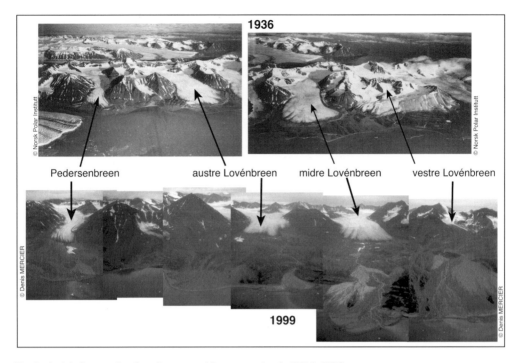

Fig. 3. Aerial photographs of northern part of Brøgger peninsula (1936–1999)

increased highly significantly by 2.8% per decade during the twentieth century (Hanssen-Bauer 2002). Statistical relationships between negative mass balance and meteorological parameters have been investigated (Lefauconnier & Hagen 1990; Hansen 1999; Lefauconnier *et al.* 1999; Nesje & Dahl 2000). Glacier retreat reflects in part a decrease in winter snowfall and a concomitant increase in rainfall. Hansen (1999) calculated the response time of midre Lovénbreen to climate change during the twentieth century to be 31 years. One consequence of these changes in climate and glacier mass budget is that seasonal meltwater discharge has been increasing (Hagen & Lefauconnier 1995). Current summer meltwater discharge amounts to about 10^7 m^3. Increased water discharge has resulted in increased sediment mobilization. Sediment transport has been estimated to be between 478 and 2009 t/km^2/a for the austre Lovénbreen catchment (Geoffray 1968; Griselin 1982) and around 940 t/km^2/a for vestre Lovénbreen catchment (Etzellmüller 2000). Paraglacial sediment reworking, particularly by meltwater runoff, is therefore increasing in importance. Changes in glacifluvial sediment input have important consequences in terms of the dynamics of the paraglacial barrier coastline (Boulton 1990; Dowdeswell & Scourse 1990),

which experiences local progradation where sediment input is augmented, but local retreat where sediment input is reduced.

Methodology

The coastal progradation analysis summarized below is part of a larger scientific analysis of the consequences of climatic change on High Arctic landscapes. Our approach combines geomorphological field observations with data obtained from satellite imagery, aerial photography and digital elevation modelling (Laffly & Mercier 1999, 2002). In the present context, the need for time-slice data at high spatial resolution dictated the employment of vertical aerial photography rather than satellite imagery. Patterns of coastline change were therefore reconstructed for 3 years for which photographic sources are available: 1966, 1990 and 1995.

The photographs were digitized to obtain a pixel size of 0.5 m, orthorectified and georeferenced. Two different resolutions were employed for the digital elevation model (DEM) to optimize results. One DEM (based on 1:10,000 maps) with a 10 m resolution was used for the main relief elements. Another, based on GPS field observations with 1 m resolution (Brossard *et al.* 1998), was used for particular landscape

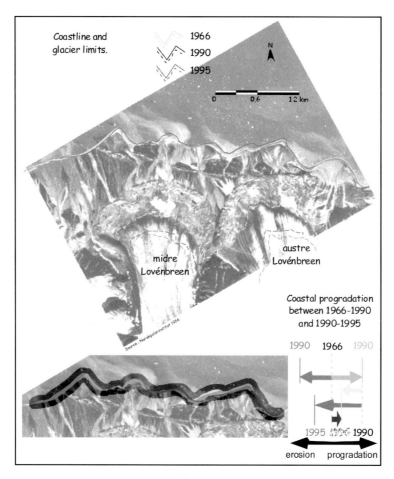

Fig. 4. Results of shoreline changes between 1966 and 1995

features, such as the Lovénbreen moraines and the coastline. For each of the above years, the images were assembled as mosaics. Different images can also be combined vertically to illustrate landscape change.

The first step in our thematic analysis consisted of a visual interpretation of shorelines and glacier limits. It subsequently proved relatively straightforward to contruct an image of shoreline change (Fig. 4). To identify progradation or erosion of the shoreline, we used the high water mark as a reference; the tidal range is about 1 m (Héquette 1986). The dominant longshore drift is northwest to southeast. The energy of tidal currents at the head of the fjord (field site) is less than in the western part of the fjord.

Results

The results of shoreline changes inferred from the airphoto sources are summarized in Fig. 4.

They show that, in sectors where meltwater sediment inputs were maintained between 1966 and 1995, shoreline progradation occurred, implying an overwhelming dominance of terrestrial sediment influx over sediment removal by wave erosion and tidal currents. Conversely, some parts of the shoreline experienced net erosion over the same time period as a consequence of loss of influx of glacifluvial sediment. There is thus a close relationship between shoreline response, the loci of glacifluvial sediment input, and paraglacial sediment reworking following glacier retreat (Mercier 2001).

For austre Lovénbreen, three distinct coastal sectors can be identified. In the eastern sector, the small sandur in front of Haavimbfjellet ceased to receive glacifluvial sediment input during the early twentieth century (river A), according to the aerial photograph of 1936. This visually confirms that this sandur has consequently undergone net erosion, averaging about

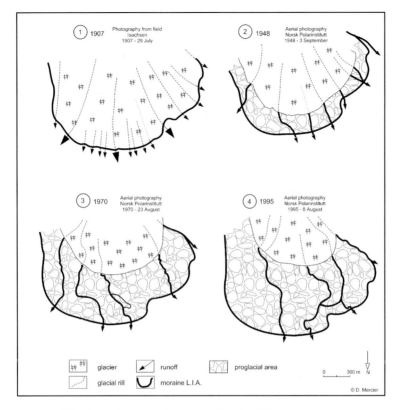

Fig. 5. River channel mobility during the twentieth century (midre Lovénbreen)

1 m/a during the study period. In contrast, the central sector (river B) prograded on average about 3 m/a between 1966 and 1995. In the western sector (river C), net shoreline progradation occurred between 1966 and 1990, but this sector is currently undergoing marked net erosion due to the termination of significant glacifluvial sediment inputs.

For midre Lovénbreen, two distinct sectors can be identified. The eastern sector (river D) experienced net stability (neither net progradation nor net retreat) between 1966 and 1995. The western sector (river E), on the other hand, experienced substantial progradation averaging around 3 m/a during the same period, accompanied by infill of the Gluudneset lagoon. These contrasts reflect river channel mobility (Fig. 5) and outlet migration across the outwash plain, sometimes produced by spontaneous channel capture and realignment of flow (Mercier 2001).

Discussion

The retreat of valley glaciers began in the early twentieth century, but initial retreat was slow:

1936 aerial photographs depict glacier margins still close to Little Ice Age terminal moraines. Glacier retreat subsequently accelerated, indicating that the glaciers were out of equilibrium with the changing climate. As the glaciers retreated, paraglacial reworking of glacigenic sediments by meltwater streams increased in importance. According to the general model of paraglacial landscape response proposed by Ballantyne (2002), this situation represents the onset of sediment reworking, when sediment availability and thus sediment flux are at their greatest.

During the period of observations (1966–1995), meltwater runoff became increasingly focused in the central part of the catchment. Drainage on the periphery of the catchments remained immaturely developed (Mercier 2001). The riverbed incised into older fluvial surface and showed the importance of downcutting taking place between the retreating glacier margin and the coast. Local slope instability in the proglacial area, related to slow summer melting of ice, represented a potential source for sediment input to the meltwater streams (Etzelmüller 2000). The progressive focusing of meltwater runoff and thus sediment discharge into the fjord has important consequences for the nearshore sediment budget and thus shoreline change. In sectors where glacifluvial sediment inputs became concentrated and maintained over the 30 year period, marked shoreline progradation has resulted. Conversely, sectors that no longer receive substantial inputs of glacifluvial sediment have experienced net erosion by wave action and tidal currents, and consequent shoreline recession. The recent evolution of the shoreline is thus intimitely related to the increasing importance of paraglacial sediment reworking, and to the changing location of meltwater systems since the end of the Little Ice Age.

Conclusions

Glaciers (austre Lovénbreen and midre Lovénbreen) have retreated more than 1 km from their Little Ice Age limits, throughout the twentieth century, and glacial meltwater has extensively reworked glacigenic sediments on exposed glacier forelands. In such areas, a paraglacial sediment transport regime has become predominant, with runoff as the dominant process.

Average shoreline progradation is estimated to amount to 3 m/a over the last 30 years, a period of uninterrupted sediment provision from the glacial runoff system.

This example shows how a polar environment responds to climatic change and how paraglacial dynamics succeed glacial dynamics in the course of landscape evolution.

This research was supported by the Centre National de la Recherche Scientifique through the Laboratory of Physical Geography in Clermont-Ferrand (UMR 6042) directed by Professor Marie-Françoise André. This research is part of a large programme supported by the French Polar Institute Paul-Emile Victor (IPEV), number 400 'geomorphoclim'.

References

BALLANTYNE, C. K. 2002. Paraglacial geomorphology. *Quaternary Science Reviews*, **21**, 1935–2017.
BOULTON, G. S. 1990. Sedimentary and sea level changes during glacial cycles and their control on glacimarine facies architecture. *In*: DOWDESWELL, J. A. & SCOURSE, J. D. (eds) *Glacimarine Environments: Processes and Sediments*, Geological Society Special Publication no. 53, 15–52.
BROSSARD, T., DESSERVY, G. & JOLY, D. 1998. Le GPS comme source de données géographiques à grande échelle. Réalisation d'un test au Svalbard. *L'Espace géographique*, **1**, 23–30.
DOWDESWELL, J. A. & SCOURSE, J. D. 1990. On the description and modeling of glacimarine sediments and sedimentation. *In*: DOWDESWELL, J. A. & SCOURSE, J. D. (eds) *Glacimarine Environments:*

Processes and Sediments, Geological Society Special Publication no. 53, 1–13.
ETZELMÜLLER, B. 2000. Quantification of thermo-erosion in pro-glacial areas – examples from Svalbard. *Zeitschrift für Geomorphologie*, **44**(3), 343–361.
FORBES, D. L. & SYVITSKI, J. P. M. 1994. Paraglacial coasts. *In*: CARTER, R. W. G. & WOODROFFE, C. D. (eds) *Coastal Evolution. Late Quaternary Shoreline Morphodynamics*, Cambridge University Press, Cambridge, 373–424.
FØRLAND, E. J. & HANSSEN-BAUER, I. 2000. Increased precipitation in the norwegian arctic: true or false? *Climatic Change*, **46**, 485–509.
GEOFFRAY, H. 1968. Etude du bilan hydrologique et de l'érosion sur un bassin partiellement englacé, Spitsberg, Baie du Roi, 79° Lat. Nord, Thesis, University of Rennes, 68 pp.
GRISELIN, M. 1982. Les modalités de l'écoulement liquide et solide sur les marges polaires (exemple du bassin Loven Est, côte NW du Spitsberg), Thesis, University of Nancy II, 500 pp.
HAGEN, J. O. & LEFAUCONNIER, B. 1995. Reconstructed runoff from the high arctic basin Bayelva based on mass balance measurements. *Nordic Hydrology*, **26**, 285–296.
HAGEN, J. O., LIESTØL, O., ROLAND, E. & JØRGENSEN T. 1993. Glacier atlas of Svalbard and Jan Mayen, Oslo, Meddelelser no. 129, 141 pp.
HAGEN, J. O., MELVOLD, K., PINGLOT, J.-F. & DOWDESWELL, J. A. 2003. On the net mass balance of the glaciers and ice caps in Svalbard, Norwegian Arctic. *Arctic, Antarctic and Alpine Research*, **35**(2), 264–270.
HANSEN, S. 1999. A photogrammetrical, climate-statistical and geomorphological approach to the post Little Ice Age changes of the Midre Lovénbreen glacier, Svalbard, Unpublished Masters thesis, University of Copenhagen–The University Courses on Svalbard (UNIS)–University of Tromsø, 103 pp.
HANSSEN-BAUER, I. 2002. Temperature and precipitation in Svalbard 1912–2050: measurements and scenarios. *Polar Record*, **38**, 225–232.
HANSSEN-BAUER, I. & FØRLAND, E. J. 1998. Long-term trends in precipitation and temperature in the Norwegian Arctic: can they be explained by changes in the atmospheric circulations patterns. *Climatic Research*, **10**, 143–153.
HÉQUETTE, A. 1986. Morpho-sédimentologie et évolution de littoraux meubles en milieu arctique, Péninsule de Brögger, Spitsberg nord-occidental, Thesis, University of Brest, 397 pp.
HUMLUM, O. 2002. Modelling late 20th-century precipitation in Nordenskiöld Land, Svalbard, by geomorphic means. *Norsk Geografisk Tidsskrift*, **56**, 96–103.
ISACHSEN, G. 1912. Exploration du Nord-Ouest du Spitsberg entreprise sous les auspices de S.A.S. le Prince de Monaco par la Mission Isachsen, Fascicule XL, Imprimerie de Monaco, 112 pp.
LAFFLY, D. & MERCIER, D. 1999. Réflexions méthodologiques sur les observations de terrain et la télédétection (cartographie des sandurs en

Baie du Roi, Spitsberg nord-occidental). *Photo-Interprétation*, **2**, 15–28, 48–58.

LAFFLY, D. & MERCIER, D. 2002. Global change and paraglacial morphodynamic modification in Svalbard. *International Journal of Remote Sensing*, **43**(21), 4743–4760.

LEFAUCONNIER, B. & HAGEN, J. O., 1990. Glaciers and climate in Svalbard: statistical analysis and reconstruction of the Brøggerbreen mass balance for the last 77 years. *Annals of Glaciology*, **14**, 148–152.

LEFAUCONNIER, B., HAGEN, J. O., ØRBÆK, J. B., MELVOLD, K. & ISAKSSON, E. 1999. Glacier balance trends in the Kongsfjorden area, western Spitsbergen, Svalbard, in relation to the climate. *Polar Research*, **18**(2), 307–313.

MERCIER, D. 2000. Du glaciaire au paraglaciaire: la métamorphose des paysages polaires au Svalbard. *Annales de Géographie*, **616**, 580–596.

MERCIER, D. 2001. *Le ruissellement au Spitsberg. Le monde polaire face aux changements climatiques*, Presses Universitaires Blaise Pascal, Clermont-Ferrand, Collection Nature & Sociétés, 278pp.

MERCIER, D. 2002. La dynamique paraglaciaire des versants du Svalbard. *Zeitschrift für Geomorphologie*, **46**(2), 203–222.

NESJE, A. & DAHL, S. O. 2000. *Glaciers and Environmental Change*, London, Arnold, 203pp.

RIPPIN, D., WILLIS, I., ARNOLD, N., HODSON, A., MOORE, J., KOHLER, J. & BJÖRNSSON, H. 2003. Changes in geometry and subglacial drainage of midre Lovénbreen, Svalbard, determined from digital elevation models. *Earth Surface Processes and Landforms*, **28**, 273–298.

SIX, D., REYNAUD, L. & LETRÉGUILLY, A. 2001. Bilans de masse des glaciers alpins et scandinaves, leurs relations avec l'oscillation du climat de l'Atlantique nord. *Comptes rendus de l'Académie des Sciences, Paris, Sciences de la terre et des planètes*, **333**, 693–698.

Holocene permafrost aggradation in Svalbard

OLE HUMLUM

Department of Physical Geography, Institute of Geosciences, University of Oslo,
PO Box 1042 Blindern, N-0316 Oslo and Department of Geology,
University Centre in Svalbard, N-9170 Longyearbyen, Norway

Abstract: The distribution and dynamics of permafrost represent a complex problem, confounded by a short research history and a limited number of deep vertical temperature profiles. This lack of knowledge is pronounced for the High Arctic, where most permafrost is found and where amplified responses to various climatic forcing mechanisms are expected. Within the High Arctic, the Svalbard region displays a unique climatic sensitivity and knowledge of Holocene, and modern permafrost dynamics in this region therefore have special interest. This paper reviews knowledge on Holocene permafrost development in Svalbard and the climatic background for this. In Svalbard, modern permafrost thickness ranges from less than 100 m near the coasts to more than 500 m in the highlands. Ground ice is present as rock glaciers, as ice-cored moraines, buried glacial ice, and in pingos and ice wedges in major valleys. Svalbard is characterized by ongoing local-scale twentieth-century permafrost aggradation, even though a distinct temperature increase around 1920 introduced relatively unfavourable climatic conditions for permafrost in Svalbard. Modern permafrost aggradation is to a large extent controlled by wind, solid precipitation and avalanche activity, and exemplifies the complexity of relating climate and permafrost dynamics.

Recent changes in the Arctic atmosphere–ice–ocean system have sparked intense discussions as to whether these changes represent episodic events or long-term shifts in the Arctic environment. Concerns about future climate change stem from the increasing concentration of greenhouse gases in the atmosphere. During the last 15 years, the Arctic has gained a prominent role in the scientific debate regarding global climatic change (Houghton *et al.* 2001). Global circulation models predict that the present and future global climatic change should be amplified in the polar regions due to feedbacks in which variations in the extent of glaciers, snow, sea ice, permafrost and atmospheric greenhouse gases play key roles. This is the basic reason for a renewed research interest in the Arctic region. Slight changes in mean annual air temperature, wind speed and precipitation have the potential to change the state of large regions of presently frozen ground (Nelson *et al.* 2001, 2002; Anisimov *et al.* 2002). Sub-continental-scale analysis of meteorological data obtained during the observation period apparently lends empirical support to the alleged high climatic sensitivity of the Arctic (Giorgi 2002). Polyakov *et al.* (2002a, b), however, recently presented updated observa-

tional trends and variations of Arctic climate and sea-ice cover during the twentieth century that question the modelled polar amplification of temperature changes registered by surface stations at lower latitudes.

There is reason therefore to evaluate past and present climate dynamics and their respective impacts on high-latitude permafrost regions such as Svalbard (Fig. 1). Permafrost is widely distributed in the Arctic and underlies as much as 20–25% of the present global land surface, affecting a wide range of ecosystems and landscapes (Péwé 1983; Zhang *et al.* 2000). In Svalbard, virtually all land areas not covered by glaciers have permafrost (Humlum *et al.* 2003). The presence of permafrost, with its overlying active layer, is a primary factor distinguishing arctic from temperate watersheds (McCann & Cogley 1972; Woo 1986; Kane & Hinzmann 1988; Woo *et al.* 1992; Brown *et al.* 2000). Permafrost is also the thermal background for a suite of permafrost-specific landforms such as ice wedges, pingos, patterned ground, palsas and rock glaciers. Massive bodies of ground ice have implications for construction work or issues relating to slope stability. From an engineering point of view, any temperature change of permafrost is likely to introduce

From: HARRIS, C. & MURTON, J. B. (eds) 2005. *Cryospheric Systems: Glaciers and Permafrost.*
Geological Society, London, Special Publications, **242**, 119–130.
0305-8719/05/$15.00 © The Geological Society of London 2005.

Fig. 1. Map of 63,000 km^2 Svalbard archipelago. About 60% is covered by glaciers (white). The remaining 25,000 km^2 (grey) are permafrost areas without permanent ice cover

changes in creep rates of existing foundations such as piles and footings, embankment foundations and variations in adfreeze bond support for pilings. Changes in the active-layer thickness will cause variations in thaw settlement during seasonal thawing, changes in frost-heave forces on pilings and total and differential frost heaving during winter (Péwé *et al.* 1981; Humlum *et al.* 2003).

At the beginning of the twenty-first century, mean annual air temperature (MAAT) is about −5°C near sea level in central Spitsbergen, and the mean annual precipitation is around 180 mm water equivalent (w.e.). The warmest period in the twentieth century was the period 1930–1940, with MAAT varying between −5 and −4°C (Førland *et al.* 1997). The average vertical precipitation gradient on Spitsbergen is 15–20% (per 100 m) in the coastal regions, while it is somewhat smaller (5–10%) in the central part of Spitsbergen (Hagen & Liestøl 1990; Hagen & Lefauconnier 1995; Killingtveit *et al.* 1996). The coastal–inland contrast in vertical precipitation gradient is assumed to be caused by enhanced orographic effects in the coastal regions, compared to inland areas (Humlum 2002). The amount and distribution of solid precipitation represents an important control on permafrost development in general (e.g., Smith & Riseborough 2002), as well as on a local scale, as will be described in the present paper.

Maximum age of Svalbard permafrost

During the Weichselian glaciation most of Svalbard was covered by glacial ice, although opinions differ as to the size and extent of the ice cover. The observation of glacial striae in several central Spitsbergen main valleys up to altitudes of about 200 m above sea level (a.s.l.; Humlum *et al.* 2003) suggests that Weichselian glaciers grew to sufficient thickness to reach the pressure melting point at the glacier sole, enabling basal sliding, even though surface air temperatures at that time were significantly below modern temperatures. Thus, during the Weichselian pre-existing permafrost in deep valleys in central Svalbard would have been exposed to thawing from both above (frictional heat generated by basal glacier sliding) and from below (geothermal heat). Adopting typical values for both frictional heat generated by glacier sliding and geothermal heat flow (e.g., Weertman 1969), the combined effect would have been thawing of at least 1 cm of permafrost per year. A typical Late Quaternary glacial period is about 100,000 years long, but only for shorter periods do glaciers grows to large size. Taking 30,000 years as an order of magnitude for the duration of extensive glaciation in the Svalbard region, this would generate a total loss of permafrost of about 200–400 m during the Weichselian. This indicates that, while permafrost in the main trunk valleys may have been eliminated completely during many Quaternary glaciations, parts of the permafrost found beneath the highest mountains only covered by relatively thin and cold-based Weichselian ice may be of considerable age, perhaps as much as 700,000 years. In addition, in areas along the west coast of Spitsbergen, which were covered by thin (or no) Weichselian glaciers, existing permafrost may also be of some antiquity, provided it survived the climatic optimum in early Holocene.

Arctic Atlantic Holocene climate

In the Northern Hemisphere, peak summer insolation occurred about 9000 years ago, when the last of the large ice sheets melted. Since then the Northern Hemisphere summers have gradually seen less incoming solar radiation. The early Holocene was relatively warm, but was also subject to a number of distinct cooling events, suggesting that the global climate may be more sensitive to solar-induced changes than is usually thought, and that such changes may represent a contributing mechanism for sub-Milankovitch climate variability (e.g., Björck et al. 2001).

Evidence from the Nordic Seas near Svalbard documents the first half of the Holocene as the warmest period during the last 13,400 years (Koç et al. 1993), followed by cooling during the Late Holocene (Koç & Jansen 1994). For the Barents Sea south of Svalbard, the early Holocene warming trend deduced from $\delta^{18}O$ records (Duplessy et al. 2001) apparently culminated between 9000 and 6500 BP, followed by rapid cooling (Ivanova et al. 2002).

In NE Greenland, 400 km west of Svalbard, Hjort (1997) documented that the main fjords were ice-free during early and mid-Holocene. Since 5700 BP, however, these fjords have mainly been ice-covered. Summarizing glaciological evidence, Kelly (1980) concluded that the Greenland Ice Sheet melted back to a position 60–120 km behind the present margin during the Early Holocene, and that a Late Holocene period of growth was initiated 3000–3500 BP, followed by major advances around 2000 BP. This development is also reflected by ice core isotope analysis from the ice sheet (Dansgaard et al. 1971, 1973). The formation of the 4208 km² Hans Tausen Ice Cap in Pearyland, northernmost Greenland, after 3900 BP (Hammer et al. 2001), as well as updated Greenland isotope series (Johnsen et al. 2001), lend support to the notion of late Holocene climatic deterioration in Greenland. Reconstructed surface temperatures from the Greenland Ice Sheet (Dahl-Jensen et al. 1998) directly testify to a 3°C net cooling during the last 4000 years.

In northern Finland, based on studies of pollen assemblages, Seppä and Birks (2001) demonstrated a marked late Holocene summer temperature decline and that the last 2000 years have been the coolest since the early Holocene. Rosén et al. (2001), from diatoms, chironomids, pollen and near-infrared spectroscopy in northern Sweden, documented late Holocene lowering of the tree-limit, caused by decreasing summer temperature. Also Hammarlund et al. (2002) reported lowering of the tree line and cooling after 3000 BP from study sites in northern Sweden. Based on botanical and glaciological evidence Nesje and Kvamme (1991) concluded that early Holocene temperatures in southern Norway were about 3°C above modern temperatures and that temperatures since 5000 BP have shown a decreasing trend.

Thus, climatic cooling have characterized the land areas around the Nordic Seas since c. 4000 BP, probably reflecting the gradual reduction of solar summer insolation caused by orbital forcing. The Nordic Seas have at the same time experienced a transition to reduced Atlantic Water influence and lowered sea surface temperatures. The Holocene transgression of the Barents Sea may have been an important driver for such major changes in oceanographic patterns, by reducing the warm Atlantic Water inflow to the Arctic Ocean (Butt et al. 2000). Atlantic Water inflows to the Arctic Ocean may also have been affected by bathymetric changes elsewhere in the Arctic (i.e., Nares Strait and other eastern Canadian Arctic Channels, and the Bering Strait). Of these, the opening of Bering Strait c. 11,000 BP might be the single most important because it introduces relatively fresh Pacific waters into the Arctic Basin. In addition, the opening of the Bering Strait may have strengthened the meridional Atlantic Water circulation through the Fram Strait, presumably at the expense of Barents Sea inflow (Forman et al. 2000).

Svalbard Holocene climate

Sea-surface and air temperatures in the Svalbard region reached their Holocene maximum 9000–7500 BP and remained above present values until c. 5000 BP (Birks 1991; Wohlfarth et al. 1993; Hjort et al. 1996). Relatively warm conditions in Svalbard during the early and mid-Holocene have been demonstrated from marine data (Salvigsen et al. 1992; Salvigsen 2002). Glaciers on Svalbard and Franz Josef Land at that time retreated to positions at or behind the present limit and remained retracted until the Late Holocene, due to high summer temperatures (Svendsen & Mangerud 1997; Lubinski et al. 1999). During the early and mid-Holocene the MAAT presumably ranged from 0 to −3°C near sea level and there was limited possibility for widespread permafrost to form or survive at such altitudes in Svalbard. Most likely, permafrost was absent or highly discontinuous near sea level. Only above 300–500 m a.s.l. did continuous permafrost persist.

The surface waters along western Svalbard have been characterized by cooling and decreased Atlantic Water influence since c. 4000 BP (e.g., Koç et al. 1993). Based on lacustrine evidence from maritime western Spitsbergen, Svendsen and Mangerud (1997) demonstrated renewed glacier growth from 4000 to 5000 BP. Additional evidence for Late Holocene glacier growth on Svalbard was presented by Snyder et al. (2000). Büdel (1977; pp. 59–60) dated a relict permafrost table at 50–60 cm depth to about 3000–3100 BP on the Hohenstaufen mountain plateau on Barentsøya (eastern Svalbard). This relict permafrost table was found 20–30 cm

Fig. 2. Pingos in upper Reindalen, central Spitsbergen, at about 80 m a.s.l. The heights of the pingos are 25–45 m; April 2001

below the modern permafrost table. Many pingos (Fig. 2) and ice wedges in Spitsbergen are found at or below the Holocene upper marine limit, believed to have been reached around 8000 BP, and must therefore postdate this. As to the timing of pingo initiation at low altitudes in central Spitsbergen, Svensson (1971) obtained a maximum age of about 2650 BP. Jeppesen (2001), working in the same area, excavated a number of ice wedges in order to study their structure and to obtain material for dating and for oxygen isotope analyses. Dating (^{14}C) suggests that these ice wedges were initiated shortly before 2900 BP, again signalling a Late Holocene establishment of permafrost near sea level in central Spitsbergen. It appears, therefore, that permafrost, pingos and ice wedges in the main valleys in central Spitsbergen are only *c.* 3000 years old. Some ice wedges found at higher altitudes may be pre-Weichselian and could have been preserved below cold-based ice (Sørbel & Tolgensbakk 2002).

Late Holocene average permafrost growth rate

Permafrost today is widely distributed on Svalbard, also at altitudes below the maximum Holocene sea level (Humlum *et al.* 2003). Much of such low-altitude permafrost is presumably of late Holocene age, for the reasons given above, at least in central Spitsbergen. In major valleys near Longyearbyen and Svea, drillings have been conducted below the upper marine limit, to map the thickness of both coal-bearing strata and permafrost in preparation for potential future mining operations (Table 1). Additional observations on permafrost thickness and temp-

erature gradients have been obtained near the former mining settlement Ny-Ålesund in NW Spitsbergen (e.g., Lauritzen 1991; Sandsbråten 1995; Haldorsen *et al.* 1996; Booij *et al.* 1998; Haldorsen & Heim 1999). Near the present coast, permafrost typically reaches a thickness of 80–100 m, while further inland, but still below an altitude of 90 m a.s.l., permafrost is typically 100–200 m thick.

Taking 100 m as a conservative thickness for low-altitude permafrost in central Spitsbergen, and assuming that this permafrost layer formed during the last *c.* 3000 years, then the average growth rate is about 3 cm/a for the Late Holocene time range considered. Most likely, initial growth rates were higher and recent rates lower than this average rate. To a first approximation, the growth of permafrost may roughly have followed the square root of time. Also hidden are the effects of Late Holocene climatic variations.

Modern Svalbard permafrost aggradation

Twentieth century climatic changes in Svalbard have been well documented by meteorological data since 1911 (Førland *et al.* 1997). Around 1920 a marked warming within 5 years changed the mean annual air temperature at sea level from about −9.5 to −4.5°C. Later, from 1957 to 1968 the air temperature dropped by about 4°C, followed by gradual increase towards the end of the twentieth century (Fig. 3). Modern MAAT at sea level is *c.* −5°C, which is about 4°C warmer compared with early twentieth-century conditions. This has not been advantageous for late twentieth-century permafrost aggradation. On this modern climatic back-

Table 1. *Measured permafrost thickness on Spitsbergen (modified from Humlum* et al. *2003)*

Location	Permafrost thickness (m)	Geothermal gradient (°C/m)	Description (m a.s.l.)	Reference
Sarkofagen	450	0.020	Mountain ridge 510	Liestøl (1976)
Liljevalchfjellet	280	0.024	Mountain ridge 580	Liestøl (1976)
Endalen	200	0.025	Valley floor 90	Liestøl (1976)
Sveagruva	125	0.050	Airstrip 30	Gregersen & Eidsmoen (1988)
	100	0.050	Beach 5	
Longyearbyen	190	0.033	Valley floor 30	Gregersen & Eidsmoen (1988)
	115	0.033	Beach 5	
Janssonhaugen	220	0.038	Mountain ridge 210	Isaksen *et al.* (2000); Isaksen (2001)
Adventdalen	107	0.04	Valley floor 40	Personal communication, Store Norske mining company (2003)
Mine 7	330	0.03	Mountain plateau 820	Author's observations (2003)

ground, it is of considerable interest that permafrost still aggrades rapidly at certain sites in the landscape, especially at the foot of talus sheets with snow avalanche activity.

Talus sheets with high snow avalanche activity in Svalbard mainly face northwest, as the dominant winter airflow across the mountains is from the southeast (Humlum 2002). At the foot of such slopes, 1–4 m avalanche snow diluted by rock debris accumulates during winter and spring (Fig. 4). During the following summer much of the snow melts, releasing incorporated rock fragments on the surface. Melt water percolates to the lower part of the snow pack before

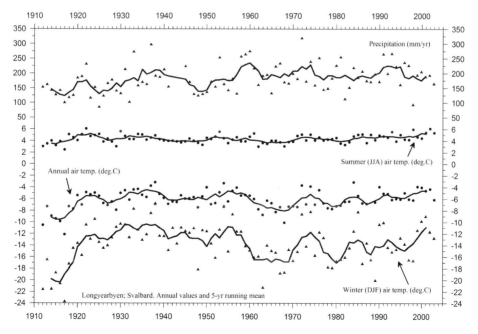

Fig. 3. Surface air temperatures and precipitation on Svalbard since initiation of measurements in 1912. Summer temperature is the average of June, July and August. Winter temperature is the average of December, January and February. Dots indicate individual years. Solid lines indicate unweighted 5 year running means

Fig. 4. Snow avalanche activity on talus sheet in upper Longyeardalen, central Spitsbergen. Avalanches are released by collapsing cornices along downwind rim of summit plateau (550 m a.s.l.). Runout zone of avalanches is about 200 m a.s.l.; 24 June 2001

refreezing, forming a basal ice layer (Woo *et al.* 1982). Part of the talus debris below the ice layer may be partly free of ice, enabling cold air to flow from the top towards the lower talus slope, as has been described from talus accumulations and rock glaciers (Keller & Gubler 1993; Harris & Pedersen 1998; Bernard *et al.* 1998; Sutter 1996; Sawada *et al.* 2003). Temperatures in the talus sheet will not rise above freezing before all avalanche snow has disappeared. As ablation consumes snow during the summer, freeing incorporated rock debris, a protective surface layer gradually forms on top of remaining avalanche snow, reducing further melting. At the end of the summer, the surface looks like the surface of a normal talus slope, but often avalanche snow is still present at 10–50 cm depth (Fig. 5). By this process, a 0–100 cm thick layer of snow and rock debris is added to the original terrain surface each year (Fig. 6). The lower part of remaining avalanche snow is typically transformed into a 5–10 cm solid ice layer by refrozen surface melt water, an analogue to the formation of superimposed ice on normal glaciers. What occurs is a net surface accumulation of snow by avalanche activity, much like the typical case for avalanche-nourished small glaciers, only in this case the ice is highly diluted by rock debris.

Thus, at such positions in the Svalbard landscape, permafrost presently aggrades 0 100 cm/a. Wind, precipitation, intensity of

bedrock weathering and snow avalanche frequency and magnitude are the main controls on this type of permafrost aggradation. Snow avalanches especially occur on downwind slopes; therefore modern permafrost aggradation on Svalbard mainly takes place along the foot of talus sheets exposed towards the northwest.

Over a series of years, a layered permafrozen deposit will accumulate, consisting of layers of rock debris and remnants of avalanche snow, each pair presumably representing the net accumulation during a single year (Fig. 6). When such an ice-rich permafrost deposit grows to a critical thickness, depending on slope, temperature and ice content, deformation commences, leading to the formation of a rock glacier. Ground-penetrating radar mapping carried out at different rock glaciers in Svalbard has demonstrated the existence of a layered internal structure much like that outlined in Fig. 6, displaying increasing up-flow dip in a down-flow direction (e.g., Berthling *et al.* 1998, 2000; Isaksen *et al.* 2000; Isaksen 2001; Korsgaard 2002). This internal structure is analogous to what since long has been described from small glaciers (e.g., McCall 1952).

Near Longyearbyen, central Spitsbergen, a meltwater channel has provided a unique means of access to the interior of a rock glacier, flowing from a segment of a larger talus sheet with high frequency of avalanches. The tunnel for about 200 m follows the interface between

Fig. 5. Late summer situation in lower part of the talus sheet shown in Fig. 4. Beneath much of the talus debris 50–100 cm thick remnants of avalanche snow are still present. At the rifle buried snow has been exposed by removal of 20–30 cm rock debris; 18 August 2001

the bedrock and the lower part of the rock glacier. The interface between bedrock and the rock glacier is sharp and the bedrock surface is rough without striae. The rock glacier body above consists of solid layers (10–100 cm thick) of relatively clean ice, interbedded with clast-supported angular rock debris layers (30–150 cm thick), with ice filling the pores (Fig. 7). Many layers were folded with their axis of folding orientated perpendicular to the rock glacier flow direction. Ice samples taken from solid ice layers in the lower part of the rock glacier typically have $\delta^{18}O$ values ranging from -10.97 to -17.47 (Humlum, unpublished data). This is identical to values obtained from snow collected at the rock glacier head upstream and from precipitation collected at nearby Longyearbyen, suggesting no major fractionation took place after burial of avalanche snow at the rock glacier head. The isotope values found also suggest that precipitation mainly took place at air temperatures between -2 and $-10°C$,

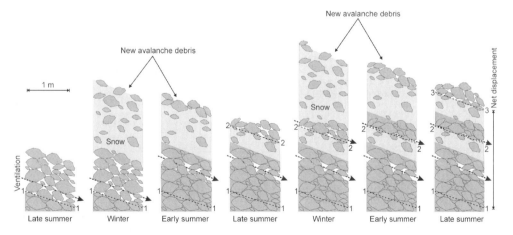

Fig. 6. Vertical displacement of active layer and permafrost table in the lower part of the talus slope with avalanche activity. Light grey indicates snow. Dark grey indicates solid ice. Numbers 1–3 indicate vertical displacement of the permafrost table. Arrows indicate internal flow of cold air

Fig. 7. Rock debris and ice in basal part of small rock glacier in upper Longyeardal, central Spitsbergen. The visible part of the rock glacier consists of layers of solid ice and layers of clast-supported angular rock debris with pore space occupied by ice. Rock glacier flow direction towards the observer; October 2002

which is comparable to modern meteorological conditions during precipitation events in the area (Humlum, unpublished data).

Conclusions

- Permafrost in Svalbard typically has a thickness of about 100–150 m near the bottom of major valleys and about 400–500 m in mountains rising above 500 m a.s.l.
- Permafrost in major valleys may have thawed below large warm-based glaciers during Quaternary glaciations, especially in central Svalbard.
- The absence of striae in the high mountains suggests cold-based Quaternary glaciers. High-altitude permafrost in Svalbard may therefore be of considerable age, perhaps as much as 700,000 years.
- Late Holocene cooling since 4000 BP has lead to widespread permafrost growth at low altitudes. Since *c.* 3000 BP at least 100 m permafrost has formed at altitudes near sea level.
- The twentieth-century climatic development has not been favourable for additional permafrost growth in Svalbard. Modern permafrost growth continues, however, at rates of 0–100 cm/year along the lower part of many talus sheets facing downwind (northwest).

- Accumulation of avalanche snow diluted by rock debris controls such downwind permafrost aggradation. This type of permafrost is therefore less dependent upon air temperature variations than permafrost in other settings, especially in low-relief areas. The geomorphic expression of this type of permafrost aggradation is talus-derived rock glaciers.

Careful critical reviews of the manuscript by Professor Chris Burn and an anonymous reviewer are highly appreciated. The Norwegian Meteorological Institute supplied meteorological data used in the investigation, and the isotope laboratory at the Department of Geophysics (Niels Bohr Institute, University of Copenhagen) carried out the $\delta^{18}O$ analyses.

References

ANISIMOV, O. A., VELICHKO, A. A., DEMCHENKO, P. F., ELISEEV, A. V., MOKHOV, I. I. & NECHAEV, V. P. 2002. Effect of climate change on permafrost in the past, present, and future. *Izvestiya Atmospheric and Oceanic Physics*, **38**, S25–S39.

BERNARD, L., SUTTER, F., HAEBERLI, W. & KELLER, F. 1998. Processes of snow/permafrost-interactions at a high-mountain site, Murteel/Corvatsch, Eastern Swiss Alps. *In: Proc. 7th International Conference on Permafrost*, Yellowknife, 35–41.

BERTHLING, I., ETZELMÜLLER, B., EIKEN, T. & SOLLID, J. L. 1998. Rock glaciers on Prins Karls Forland, Svalbard. I: Internal structure, flow velocity and morphology. *Permafrost and Periglacial Processes*, **9** 135–145.

BERTHLING, I., ETZELMÜLLER, B., ISAKSEN, K. & SOLLID, J. L. 2000. Rock glaciers on Prins Karls Forland. II: GPR soundings and the development of internal structures. *Permafrost and Periglacial Processes*, **11**, 157–169.

BIRKS, H. H., 1991. Holocene vegetational history and climatic change in West Spitsbergen–plant macrofossils from Skardtjørna. *The Holocene*, **1**, 209–218.

BJÖRCK, S., MUSCHELER, R., KROMER, B., ANDRESEN, C. S., HEINEMEIER, J., JOHNSEN, S. J., CONLEY, D., KOÇ, N., SPURK, M. & VESKI, S. 2001. High-resolution analyses of an early Holocene climate event may imply decreased solar forcing as an important climate trigger. *Geology*, **29**, 1107–1110.

BOOIJ, M., LEIJNSE, A., HALDORSEN, S., HEIM, M. & RUESLÅTTEN, H. 1998. Subpermafrost groundwater modelling in Ny-Ålesund. *Nordic Hydrology*, **29**, 385–396.

BROWN, J., HINKEL, K.M. & NELSON, F.E. 2000. The circumpolar active layer monitoring (CALM) program: Research designs and initial results. *Polar Geography*, **24**, 163–258.

BÜDEL, J. 1977. *Klima-Geomorphologie*. Gebrüeder Borntraeger, Berlin.

BUTT, F. A., ELVERHØI, A., SOLHEIM, A. & FORSBERG, C. F. 2000. Deciphering late Cenozoic evolution of the western Svalbard Margin based of ODP Site 986 results. *Marine Geology*, **169**, 373–390.

DAHL-JENSEN, D., MOSEGAARD, K., GUNDESTRUP, N., CLOW, G. D., JOHNSEN, S. J., HANSEN, A.W. & BALLING, N. 1998. Past temperatures directly from the Greenland Ice Sheet. *Science*, **282**, 268–271.

DANSGAARD, W., JOHNSEN, S. J., CLAUSEN, H. B. & LANGWAY, C. C. JR. 1971. Climatic record revealed by the Camp Century ice core. *In*: TUREKIAN, K. K. (ed.) *The Late Cenozoic Glacial Ages*. Yale University Press, New Haven, CT, 37–56.

DANSGAARD, W., JOHNSEN, S. J., CLAUSEN, S. J. & GUNDESTRUP, N. 1973. Stable isotope glaciology. *Meddelelser om Grønland*, **197**, 1–53.

DUPLESSY, J. C., IVANOVA, E. V., MURDMAA, I. O., PATERNE, M. & LABEYRIE, L. 2001. Holocene paleoceanography of the Northern Barents Sea and variations of the northward heat transport by the Atlantic Ocean. *Boreas*, **30**, 2–16.

FORMAN, S. L., MASLOWSKI, W., ANDREWS, J. T., LUBINSKI, D. J., STEELE, M., ZHANG, Y. & LAMMERS, R. 2000. Researchers explore Arctic freshwater's role in ocean circulation. *EOS, Transactions, American Geophysical Union*, **81**, 169–174.

FØRLAND, E. J., HANSSEN-BAUER, I. & NORDLI, P. Ø. 1997. *Climate Statistics & Long Term Series of Temperature and Precipitation at Svalbard and Jan Mayen*. Det Norske Meteorologiske Institutt, Report no. 21/97 klima.

GIORGI, F. 2002. Variability and trends of sub-continental scale surface climate in the twentieth century. Part I: Observations. *Climate Dynamics*, **18**, 675–691.

GREGERSEN, O. and EIDSMOEN, T. 1988. Permafrost conditions in the shore area of Svalbard. Norwegian Geotechnical Institute, publication 174, Oslo, Norway.

HAGEN, J. O. & LEFAUCONNIER, B. 1995. Reconstructed runoff from the High Arctic basin Bayelva based on mass-balance Measurements. *Nordic Hydrology*, **26**(4/5), 285–296.

HAGEN, J. O. & LIESTØL, O. 1990. Long-term glacier mass-balance investigations in Svalbard, 1950–88. *Annals of Glaciology*, **14**, 102–106.

HALDORSEN, S. & HEIM, M. 1999. An Arctic groundwater system and its dependence upon climate change: An example from Svalbard. *Permafrost and Periglacial Processes*, **10**, 137–149.

HALDORSEN, S., HEIM, M. & LAURITZEN, S.-E. 1996. Subpermafrost Groundwater, western Svalbard. *Nordic Hydrology*, **27**, 57–68.

HAMMARLUND, D., BARNEKOW, L., BIRKS, H. J. B., BURCHARD, B. & EDWARDS, T. W. D. 2002. Holocene changes in atmospheric circulation recorded in the oxygen-isotope stratigraphy of lacustrine carbonates from northern Sweden. *The Holocene*, **12**, 339–251.

HAMMER, C. U., JOHNSEN, S. J., CLAUSEN, H. B., DAHL-JENSEN, D., GUNDESTRUP, N. & STEFFENSEN, J. P. 2001. The Paleoclimatic Record from a 345 m long Ice Core from the Hans Tausen Iskappe. Copenhagen, Danish Polar Center. *Meddelelser om Grønland Geoscience*, **39**, 87–95.

HARRIS, S. A. & PEDERSEN, D. E. 1998. Thermal regimes beneath coarse blocky materials. *Permafrost and Periglacial Processes*, **9**, 107–120.

HJORT, C. 1997. Glaciation, climate history, changing marine levels and the evolution of the Northeast Water Polynia. *Journal of Marine Systems*, **10**, 23–33.

HJORT, C., MANGERUD, J., ANDRIELSSON, L., BONDEVIK, S., LANDVIK, J. Y. & SALVIGSEN, O. 1996. Radiocarbon dated common mussel *Mytilus edulis* from eastern Svalbard and the Holocene marine climatic optimum. *Polar Research*, **14**, 239–243.

HOUGHTON, J. T., DING, Y., GRIGGS, D. J., NOUGER, M., VAN DER LINDEN, P. J., DAI, X., MASKELL, K and JOHNSON C. A. 2001. Climate change 2001: The Scientific Basis. Contribution of Working Group I to the Third Assessment Report of the Intergovernmental Panel on Climate Change, Cambridge University Press, 881pp.

HUMLUM, O. 2002. Modelling late 20th century precipitation in Nordenskiöld Land, central Spitsbergen, Svalbard, by geomorphic means. *Norwegian Geographical Journal*, **56**, 96–103.

HUMLUM, O., INSTANES, A. & SOLLID, J. L. 2003. Permafrost in Svalbard; a review of research history, climatic background and engineering challenges. *Polar Research*, **22**(2), 191–215.

ISAKSEN, K. 2001. Past and present ground thermal regime, distribution and creep of permafrost – case studies in Svalbard, Sweden and Norway.

Thesis, Faculty of Mathematics and Natural Sciences, University of Oslo.

ISAKSEN, K., ØDEGÅRD, R. S., EIKEN, T. & SOLLID, J. L. 2000. Composition, flow and development of two tongue-shaped rock glaciers in the permafrost of Svalbard. *Permafrost and Periglacial Processes*, **11**, 241–257.

IVANOVA, E. V., MURDMAA, I. O., DUPLESSY, J.-C. & PATERNE, M. 2002. Late Weichselian to Holocene paleoenvironments in the Barents Sea. *Global and Planetary Change*, **34**, 209–218.

JEPPESEN, J. W. 2001. Palæoklimatiske indicatorer for central Spitsbergen, Svalbard. Eksemplificeret ved studier af iskiler og deres værtssedimenter. M.Sc. thesis, The University Courses on Svalbard (UNIS).

JOHNSEN, S. J., DAHL JENSEN, D., GUNDESTRUP, N., STEFFENSEN, P., CLAUSEN, H. B., MILLER, H., MASSON-DELMOTTE, V., SVEINBJÖRNSDOTTIR, A. E. & WHITE, J. 2001. Oxygen isotope and palaeotemperature records from six Greenland ice-core stations: Camp Century, Dye-3, GRIP, GISP2, Renland and North GRIP. *Journal of Quaternary Science*, **16**, 299–307.

KANE, D. L. & HINZMAN, L. D., 1988. Permafrost hydrology of a small arctic watershed. *In*: SENNESET, K. (ed.), *Permafrost: Fifth International Conference*, Trondheim, Norway. Tapir Press, 590–595.

KELLER, F. & GUBLER, H. U. 1993. Interaction between snow cover and high mountain permafrost at Murtel/Corvatsch, Swiss Alps. *In*: *Sixth International Conference on Permafrost*, Beijing, Vol. 1, 332–337.

KELLY, M. 1980. *The Status of the Neoglacial in Western Greenland*. The Geological Survey of Greenland, Report no. 96

KILLINGTVEIT, Å., PETTERSON, L. E. & SAND, K. 1996. Water balance studies at Spitsbergen, Svalbard. *In*: *International Northern Research Basins Symposium and Workshop*, October 1994. Spitsbergen, Trondheim, 77–94.

KOÇ, N. & JANSEN, J. 1994. Response of the high-latitude Northern Hemisphere to orbital climate forcing: Evidence from the Nordic Seas. *Geology*, **22**, 523–526.

KOÇ, N., JANSEN, E. & HAFLIDASON, H. 1993. Paleoceanographic reconstructions of surface ocean conditions in the Greenland, Iceland and Norwegian seas through the last 14ka based on diatoms. *Quaternary Science Reviews* **12**, 115–140.

KORSGAARD, S. 2002. Geomorfologisk analyse af Ugledalen blokgletscher. Det Centrale Spitsbergen, Svalbard. M.Sc. thesis, The University Courses on Svalbard (UNIS).

LAURITZEN, S.-E. 1991. Groundwater in cold climates: interaction between glacier and karst aquifers. *In*: GJESSING, Y., HAGEN, J.O., SAND, K. & WOLD, B. (eds) *Arctic Hydrology, Present and Future Tasks* Norwegian National Committee for Hydrology, Oslo, Report 23, 139–146.

LIESTØL, O. 1976. Pingos, springs, and permafrost in Spilsbergen. Notsk Polarinstitutt Årbok, 1975, 7–29.

LUBINSKI, D. J., FORMAN, S. L. & MILLER, G. H., 1999. Holocene glacier and climate fluctuations on Franz Josef Land, Arctic Russia, 80°N. *Quaternary Science Reviews*, **18**, 85–108.

McCALL, J. G. 1952. The internal structure of a cirque glacier. *Journal of Glaciology*, **2**, 122–130.

McCANN, S. B. & COGLEY, J. G., 1972. Hydrological observations on a small arctic catchment, Devon Island. *Canadian Journal of Earth Sciences*, **9**, 361–365.

NELSON, F. E., ANISIMOV, O. A. & SHIKLOMANOV, N. I. 2002. Climate change and hazard zonation in the circum-Arctic permafrost regions. *Natural Hazards*, **26**, 203–225.

NELSON, F. E., ANISIMOV, O. A. & SHIKLOMANOV, N. I. 2001. Subsidence risk from thawing permafrost – the threat to man-made structures across regions in the far north can be monitored. *Nature*, **410**, 889–890.

NESJE, A. & KVAMME, M. 1991. Holocene glacier and climate variations in western Norway: Evidence for early Holocene glacier demise and multiple Neoglacial events. *Geology*, **19**, 610–612.

PÉWÉ, T. 1983. Alpine permafrost in the contiguous United States: A review. *Arctic and Alpine Research*, **15**, 145–156.

PÉWÉ, T., ROWAN, D. E. & PÉWÉ, R. H. 1981. Engineering geology of the Svea lowland, Spitsbergen, Svalbard. *Frost i jord*, **23**, 3–11.

POLYAKOV, I., AKASOFU, S.-I., BHATT, U., COLONY, R., IKEDA, M., MAKSHTAS, A., SWINGLEY, C., WALSH, D. & WALSH, J. 2002a. Trends and variations in Arctic climate systems. *Transactions, American Geophysical Union (EOS)*, **83**, 547–548.

POLYAKOV, I. V., ALEKSEEV, G. V., BEKRYAEV, R. V., BHATT, U., COLONY, R. L., JOHNSON, M. A., KARKLIN, V. P., MAKSHTAS, A. P., WALSH, D. & YULIN, A. V. 2002b. Observationally based assessment of polar amplification of global warming. *Geophysical Research Letters*, **29**(18), 1878, 25-1–25-4.

ROSÉN, P., SEGERSTRÖM, U., ERIKSSON, L., RENBERG, I. & BIRKS, H. J. B. 2001. Holocene climatic change reconstructed from diatoms, chironomids, pollen and near-infrared spectroscopy at an alpine lake (Sjuodjijaure) in northern Sweden. *The Holocene*, **11**, 551–562.

SALVIGSEN, O. 2002. Radiocarbon-dated *Mytilus edulis* and *Modiolus modiolus* from northern Svalbard: Climatic implications. *Norwegian Journal of Geography*, **56**, 56–61.

SALVIGSEN, O., FORMAN, S. L. & MILLER, G. H. 1992. Thermophilous molluscs on Svalbard during the Holocene and their paleoclimatic implications. *Polar Research*, **11**, 1–10.

SANDSBRÅTEN, K. 1995. Vannbalanse i et lite, arktisk nedbørsfelt. Tvillingvatn, Svalbard. Rapportserie: Hydrology, University of Oslo, no. 43.

SAWADA, Y., ISHIKAWA, M. & ONO, Y. 2003. Thermal regime of sporadic permafrost in a block slope on

Mt. Nishi-Nupukaushinupuri, Hokkaido Island, Northern Japan. *Geomorphology*, **52**, 121–130.

SEPPÄ, H. & BIRKS, H. J. B. 2001. July mean temperature and annual precipitation trends during the Holocene in the Fennoscandian tree-line area: Pollen-based climate reconstruction. *The Holocene*, **11**, 527–539.

SMITH, M. W. & RISEBOROUGH, D. W. 2002. Climate and the Limits of Permafrost: A Zonal Analysis. *Permafrost and Periglacial Processes*, **13**, 1–15.

SNYDER, J. A., WERNER, A. & MILLER, G. H. 2000. Holocene cirque glacier activity in western Spitsbergen, Svalbard; sediment records from Proglacial Linnévatnet. *The Holocene*, **10**, 555–563.

SØRBEL, L. & TOLGENSBAKK, J. 2002. Ice-wedge polygons and solifluction in the Adventdalen area, Spitsbergen, Svalbard. *Norsk Geografisk Tidsskrift*, **56**, 62–66.

SUTTER, F. 1996. Untersuchung von Schloten in der Schneedecke des Blokgletscher Murtel am Corvatsch. M.Sc. thesis, Department of Zurich.

SVENDSEN, J. I. & MANGERUD, J., 1997. Holocene glacial and climatic variations on Spitsbergen, Svalbard. *The Holocene*, **7**, 45–57.

SVENSSON, H. 1971. Pingos i yttre delen av Adventdalen. *Norsk Polarinstitutt Årbok*, **1969**, 168–174.

WEERTMAN J. 1969. Water lubrication mechanism of glacier surges. *Canadian Journal of Earth Sciences*, **6**, 929–939.

WOHLFARTH, B., LEMDAHL, G., OLSSON, S., PERSSON, T., SNOWBALL, I., ISING, J. & JONES, V., 1993. Early Holocene environment on Bjørnøya (Svalbard) inferred from multidisciplinary lake sediment studies. *Polar Research*, **14**, 253–275.

WOO, M.-K. 1986. Permafrost hydrology in North America. *Atmosphere and Ocean*, **24**, 201–234.

WOO, M.-K., HERON, R. & MARSH, P. 1982. Basal ice in high arctic snow packs. *Arctic and Alpine Research*, **14**, 251–260.

WOO, M.-K., LEWKOWICZ, A. G. & ROUSE, W. R. 1992. Response of the Canadian permafrost environment to climatic change. *Physical Geography*, **13**, 287–317.

ZHANG, T., HEGINBOTTOM, J. A., BARRY, R. E. & BROWN, J. 2000. Further statistics on the distribution of permafrost and ground ice in the Northern Hemisphere. *Polar Geography*, **24**, 126–131.

Experimental simulation of ice-wedge casting: processes, products and palaeoenvironmental significance

CHARLES HARRIS[1] & JULIAN B. MURTON[2]

[1]*School of Earth, Ocean and Planetary Sciences, Cardiff University, Cardiff CF10 3YE, UK*
(e-mail: harrisc@cardiff.ac.uk)
[2]*Department of Geography, University of Sussex, Brighton BN1 9QJ, UK*

Abstract: In six experiments, model ice wedges 150 mm high and ≤ 50 mm wide were thawed from the surface downward at $30g$ in a geotechnical centrifuge, simulating ice-wedge casting during progressive active-layer deepening through permafrost. The frozen host soils ranged from sand to silt to clayey silt. Resulting ice-wedge pseudomorphs indicated that the degree of deformation during casting was determined by factors that control thaw consolidation. Hence, deformation increased as the host sediments became finer-grained and more ice-rich. Thaw of host sand at 15 and 20% gravimetric ice content and host silt at 20 and 40% ice content resulted in the formation of partially-developed ice-wedge pseudomorphs with an upper 'plug' of sediment derived largely from the cover soil, a central tunnel and a basal plug of sediment. Thaw of host clayey silt at 30 and 60% ice content resulted in the formation of fully developed pseudomorphs that were significantly narrower and shorter than the initial ice wedges. The experiments support the hypothesis that ice-wedge pseudomorphs tend to be better preserved in coarser-grained sediments that are not ice-rich and therefore deform little during thaw of constituent ice wedges. This selective preservation has probably led to underestimation of ice-wedge pseudomorphs in fine-grained soils that were originally ice-rich and therefore a bias towards reconstructing cold temperatures.

Ice-wedge pseudomorphs provide unambiguous evidence for the former occurrence of permafrost. Their palaeoenvironmental significance, however, is complicated by uncertainties about their value as palaeotemperature indicators (Murton & Kolstrup 2003) and about their selective preservation in the geological record. According to Black (1976), preservation of ice-wedge pseudomorphs is facilitated in coarse-grained sediments such as gravel because they tend to contain little ground ice and therefore deform little on thaw. Conversely, fine-grained sediments, which are often supersaturated with ground ice, should rarely leave pseudomorphs if thaw occurs rapidly enough to allow flowage of the host material into the void left by the thawing ice wedge (cf. Dylik 1966). Limited field evidence consistent with selective preservation was provided by observations of pseudomorphs above partially thawed ice wedges in ice-poor sand but not in ice-rich silt and clay in the Richards Island area of the western Arctic coast, Canada (Murton & French 1993). However, rigorous testing of this hypothesis remains to be undertaken.

This paper presents results from laboratory experiments designed to test the hypothesis of selective preservation of ice-wedge pseudomorphs and to investigate the mechanisms of casting and the types of resulting sedimentary structures. Six experiments were carried out under an enhanced gravitational field in a geotechnical centrifuge. Downward thaw of scaled models of permafrost containing an ice wedge of uniform size was conducted in sand, silt and a silt–clay mixture at gravimetric ice contents of 15–60%. We present data on soil temperature and porewater pressure during thaw, and describe the resulting ice-wedge pseudomorphs. This allows discussion of the role of thaw consolidation in determining the mechanisms and degree of casting. We then compare the artificial pseudomorphs with natural ones and discuss their palaeoenvironmental significance.

Ice wedges and ice-wedge casting

Of the three types of ice wedges distinguished by Mackay (1990), epigenetic wedges are the most commonly reported in the literature. Such

From: HARRIS, C. & MURTON, J. B. (eds) 2005. *Cryospheric Systems: Glaciers and Permafrost.*
Geological Society, London, Special Publications, **242**, 131–143.
0305-8719/05/$15.00 © The Geological Society of London 2005.

wedges grow in pre-existing permafrost and are usually much younger than the host material. Epigenetic ice wedges generally have a maximum width, normal to their axial plane, of c. 1.0–1.5 m, a depth of 3.0–4.0 m and form a polygonal network (French 1996, pp. 89–90). Ice wedges have been observed in sediments with a wide variety of textures and ice contents.

Thaw of ice wedges in flattish terrain often results from a general deepening of the active layer or from localized thaw beneath standing water in ice-wedge troughs. Thaw of excess ice and seepage of resulting meltwater results in thaw consolidation, which is usually expressed by thermokarst subsidence of the ground surface above thawing ice wedges. In contrast, ice wedges on hillslopes and near bluffs often melt by thermal erosion, as water seeps along ice-wedge troughs or in tunnels through the underlying wedges. The voids formed by thawing of the ice wedge may infill with material from above or beside the ice-wedge, a process known as ice-wedge casting.

Casting mechanisms and their controlling factors are difficult to monitor in the field and so are inferred largely from Pleistocene ice-wedge pseudomorphs in mid-latitudes (Dylik 1966; Vandenberghe 1983), supplemented by descriptions of partially thawed ice wedges in contemporary permafrost environments (Harry & Gozdzik 1988; Murton & French 1993) and monitoring the thaw of artificial ice wedges in the laboratory (Murton & Harris 2003). Casting may proceed progressively as ice wedges thaw, for example during permafrost degradation, or in a complex manner, interrupted by intervening stages of permafrost aggradation (Mackay 1988). However, in either case, seasonal active-layer refreezing complicates permafrost thaw.

From these considerations we have developed a simple model that simulates casting of epigenetic ice wedges during progressive deepening of an active layer beneath flat ground. We do not, however, include the effects of seasonal active-layer freezing and thawing that is superimposed on any progressive lowering of the permafrost table in the field setting. Frozen host sediments of different texture and ice contents were simulated in order to examine their effects on the mechanisms of ice-wedge casting and on the size and shape of the resulting casts.

Experimental methodology

The experimental methodology is described by Murton & Harris (2003) and is based on geotechnical centrifuge physical modelling of downward thaw of ice-rich permafrost. This approach allows correct self-weight stresses to be generated within scaled models by compensating for the reduced model linear dimensions (1/N scale) by a corresponding increase in gravitational acceleration (Ng). It has been shown that scaling conflicts do not arise in centrifuge modelling of thaw consolidation since both time for conductive heat transfer and time for seepage force similarity scale at $1/N^2$ in the model (Coce et al. 1985; Savidou 1988) and the thaw consolidation ratio is therefore the same at model and prototype scales (Harris et al. 2001).

Models were all constructed at 1/30 scale in a polypropylene test box of internal dimensions 750 mm long by 450 mm wide by 500 mm deep. In models 1–3, five successive c. 30 mm thick layers of soil premixed to a known moisture content were laid above a basal sand layer. In models 4–6, only four layers of soil, each c. 37 mm in thickness, were laid above the basal sand layer. Each layer was frozen before the next was added. Successive soil layers were separated by 1–2 mm thick marker horizons of soil with a contrasting colour or texture in order to highlight any deformation structures formed during thaw. The soil layers and markers were laid on either side of a wedge-shaped aluminium mould (Fig. 1).

After the uppermost soil layer had been frozen, the mould was removed from the soil, leaving a wedge-shaped trough within frozen soil into which a model ice wedge was inserted. The wedge was 150 mm high and tapered uniformly downward from a maximum width of 50 mm (Fig. 2). Any gaps between the sides of the wedge and trough, usually no more than 1–2 mm wide, were filled with chilled water and frozen. A model active layer 25 mm thick was placed above the wedge and moistened with warm water in order to expedite thaw of the model.

Six miniature Druck PDCR-81 pore pressure transducers and up to 14 Type K thermocouples were buried in the soil during model construction (Fig. 2). The transducers were placed at 50 and 100 mm distances from the edge of the wedge in order to determine (1) if excess pore pressures resulted during thaw and (2) whether these were associated with lateral or vertical hydraulic gradients. Monitoring of pore pressure also facilitates interpretation of soil rheology and hence casting mechanisms. The thermocouples permitted monitoring of thaw rate and isotherm patterns adjacent to the wedge. Temperature and pore pressure were recorded at 10 s intervals.

The models were thawed from the surface downward under an acceleration of 30g in the

Fig. 1. Construction of permafrost and ice-wedge model. Note that each layer of host soil was frozen prior to placing the next marker soil layer and horizon. The aluminium mould used to form the ice-wedge void was frozen into the soil. When preparation of the frozen host soil layers was completed, the mould was removed and a frozen ice wedge of identical dimensions was placed into the void to fill it (see text)

Cardiff Geotechnical Centrifuge Facility. Thus, dimensional scaling during thaw between model and prototype was ×30. The model ice wedges were equivalent to prototype-scale wedges 4.5 m high and ≤1.5 m wide, the active layer thickness was equivalent to 0.75 m and the total thickness of frozen host soil (i.e., permafrost) prior to commencement of thawing was *c.* 5 m.

Six models were constructed to 1/30 scale (Table 1). Host sediments comprised medium sand (models 1 and 2), natural loessic silt collected from Pegwell Bay, Kent, UK (models 3 and 4) and mixtures of Pegwell silt and kaolinite clay (models 5 and 6; Fig. 3). Gravimetric moisture contents were *c.* 15–60%. Models 3 and 6 had moisture contents above the liquid limit, model 4 had a moisture content approximately equal to the liquid limit, and model 5 had a moisture content between the plastic limit and the liquid limit.

Centrifuge tests lasted between *c.* 8 h (model 1, scaling to a prototype time of 300 days) and 22.5 h (model 6, scaling to a prototype time of

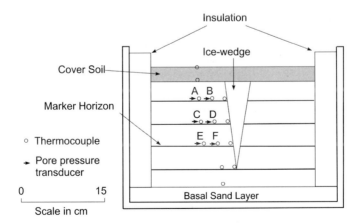

Fig. 2. Vertical section through ice-wedge model 6, showing location of porewater pressure transducers and thermocouples adjacent to the ice wedge, and of marker horizons in the host silt–clay soil

Table 1. *Ice-wedge model parameters*

Model	Grain size	Moisture content[a] (%)	Host soil liquid limit (%)	Plastic limit (%)	Cover soil	Cover soil liquid limit (%)	Plastic limit (%)	Average thaw rate prototype timescale (mm/day)	Average thaw settlement (%)
1	Medium sand	15	—	—	Sand	—	—	14.5	5
2	Medium sand	20	—	—	Sand	—	—	13.9	6
3	Pegwell Bay silt	40	20	13	Prawle silt	34	13	8.9	30
4	Pegwell Bay silt	20	20	13	Prawle silt	34	13	8.5	13
5	$\frac{2}{3}$ silt, $\frac{1}{3}$ clay[b]	30	33	18	Glacial Silt	39	14	7.8	17
6	$\frac{2}{3}$ silt, $\frac{1}{3}$ clay[b]	60	33	18	Glacial Silt	39	14	6.9	44

[a]Approximate gravimetric moisture content. [b]Kaolinite.

844 days). Soil drainage was allowed during thaw, except in model 2. Time-lapse photographs were taken through the clear Perspex front of the strongbox containing the test box, and videos were recorded through the front and top of the model to document ice-wedge casting. After thawing, the models were allowed to drain for 2–4 weeks before they were vertically sectioned at 20–25 mm increments and structures within them were measured, sketched and photographed.

Results

Temperature

In all tests the model surface was exposed to the ambient air temperature of the centrifuge test chamber (*c.* 17°C), which resulted in soil surface temperatures increasing to between 10 and 15°C through each test. Thaw rates within the host sediment were highest in models 1 and 2 (prototype scale, 14.5 and 13.9 mm/day, respectively) and lowest in models 5 and 6 (prototype scale, 7.8 and 6.9 mm/day, respectively; Table 1). The rate of penetration of the thaw front is partly a function of the latent heat necessary to thaw soil ice, and thaw rates within models 3–6 (Pegwell Bay silt and the silt/clay mix) decreased progressively with increasing ice content.

Since the latent heat required to melt the ice wedge in all models exceeded that necessary to melt the frozen host sediments, thawing was delayed immediately adjacent to the wedge compared with distal locations within the host soils (Fig. 4). The contrast in latent heat needed to

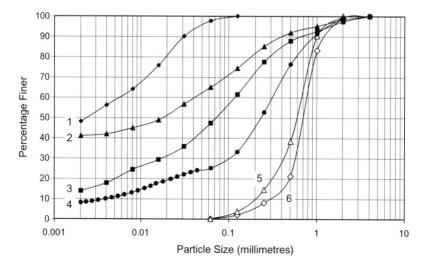

Fig. 3. Grain-size distributions of test soils: (1) cover soil, models 5 and 6; (2) host soil, models 5 and 6; (3) host soil, models 3 and 4; (4) cover soil, models 3 and 4; (5) host soil, models 1 and 2; (6) cover soil, models 1 and 2

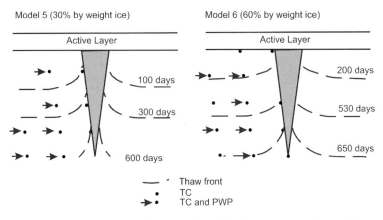

Fig. 4. Thaw front penetration, models 5 and 6, showing effect of moisture contents of host soils

thaw the host soil and that needed to thaw the ice wedge is greater at lower soil moisture contents so that, in these experiments, the differential rate of thaw penetration between the wedge and soil increased as ice contents decreased.

Thaw settlement

Consolidation of the host soil during and after thawing leads to a reduction in its thickness, termed thaw settlement. Thaw settlement is a function of initial ice content of the soil and its coefficient of volume compressibility. Sands are non-compressible and non-frost-susceptible, so that thaw settlement is likely to be minimal. Silts and clays are, however, both compressible and frost susceptible, and thaw settlement may be large, depending on the initial excess ice content. Average thaw settlement is given in Table 1 for each model. Clearly models 1 and 2 showed little settlement, while in models 3–6 settlement was greater, the ice content being the main determining factor.

Porewater pressures

Positive pore pressures were not recorded in sand model 1, and pressures were hydrostatic within the undrained model 2. This was anticipated due to low ice contents, little thaw settlement and high permeability of these models. In both the Pegwell Bay silt models (models 3 and 4) pore pressures were positive, but below hydro-static during thaw, with highest values recorded in model 3 (40% gravimetric moisture content, Fig. 5). During thaw of the Pegwell Bay silt models, therefore, thaw consolidation did not generate pore pressures sufficiently high to

reduce the frictional strength of the host soil sufficiently to promote deformation adjacent to the thawing ice wedge.

Similarly, in model 5 (Pegwell Bay silt + 30% kaolinite clay, 30% gravimetric moisture), pore pressures were positive but below hydrostatic. In contrast, thaw of model 6 (silt/clay mixture with 60% gravimetric moisture content) was associated with thaw settlement of around 44% of the initial frozen thickness and high pore pressures (Fig. 6). Thus, although pore pressures approximated hydrostatic relative to transducer depths in the frozen model, at the time that the thaw plane reached each transducer and there-after, recorded pore pressures were nearer geo-static than hydrostatic as a result of (a) low initial soil density and (b) reduction in overbur-den thickness as thaw consolidation proceeded. Pore pressures in the upper part of the model fell only slightly through the thaw phase, reflecting continued upward migration of meltwater from below. Pore pressures at greater depths also remained high during the period when the adjacent ice wedge thawed, so that frictional strength of the host soil was low, facilitating deformation of the void left by the melting ice wedge.

Sedimentary structures

A variety of sedimentary structures formed during thaw of the model ice wedges. The main group of structures are more accurately termed 'ice-wedge pseudomorphs' rather than 'ice-wedge casts'. The latter implies that the size and shape of the initial ice wedge are preserved exactly (Harry & Gozdzik 1988), unlike the structures formed in the models. In models 1–4, infilling and therefore casting were incom-plete. A surface trough was underlain by a

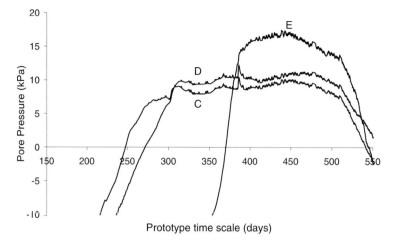

Fig. 5. Pore pressures recorded during thaw of model 3. For transducer locations see Fig. 2. Note that negative pressures (suctions) were recorded at each transducer while the surrounding soil was frozen, but pressures rapidly became positive as the thaw front reached each transducer

partially developed ice-wedge pseudomorph with a tripartite structure: (1) an upper 'plug' of sediment derived from the active layer cover soil above; (2) a central tunnel; and (3) a basal plug of sediment that formed a 'toe' to the pseudomorph. In models 5 and 6 the size and shape of the infills differed significantly from those of the initial ice wedge. These structures are illustrated in Fig. 7 and their dimensions at model and prototype scale are listed in Table 2. The main structures observed in each model are discussed below.

Model 1 (Fig. 7a). The surface trough had a mean width of 89 ± 8 mm (2.67 ± 0.24 m at prototype scale) and a mean depth of 35 ± 3 mm (1.05 ± 0.09 m, prototype). Beneath the trough was a sand plug infilling the upper third to half of the volume previously filled by the ice wedge. The sediment comprised the original cover soil, except locally where small amounts of host sediment occurred along one or both sides of the plug. The mean maximum width of the plug was 41 ± 2 mm (1.23 ± 0.06 m prototype dimension) and the mean height was

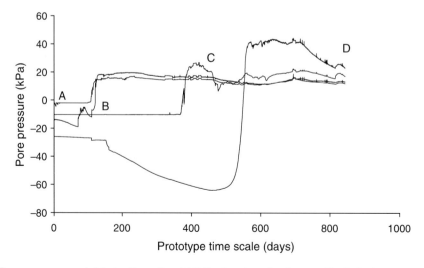

Fig. 6. Pore pressures recorded during thaw of model 6. For transducer locations see Fig. 2. Again pore pressures were negative when frozen, but rose rapidly as the thaw front reached each transducer

48 ± 3 mm (1.44 ± 0.09 m, prototype). In vertical section, the shape of the plug varied from bulbous with convex-outward sides to downward tapering with linear sides and a flat bottom. The sides of the plug varied from sharp (where marker layers adjacent to the plug showed little or no downturning) to diffuse (where there was significant down turning of markers). Within the upper centre of the plug was a poorly defined involution in the form of a downward lobate protrusion.

A tunnel beneath the plug had a mean maximum width of 32 ± 3 mm (0.96 ± 0.09 m, prototype) and a mean height of 63 ± 5 mm (1.89 + 0.15 m, prototype). The tunnel sides were generally linear, converging with depth, and marking the sides of the former ice wedge.

Marker layers adjacent to the tunnel were mostly horizontal. The tunnel roof was generally arched or flat, and the floor commonly comprised a broad central ridge flanked by narrow furrows.

A small basal plug of sand (toe) occupied the space previously filled by the toe of the ice wedge. The plug was poorly defined with gradational sides and appeared to have a U-shaped base. The infill was loosely packed. Marker layers adjacent to the plug were usually smoothly downturned.

Model 2 (Fig. 7b). The structures observed in model 2 were broadly similar in size, shape and composition to those in model 1 (Table 2). The main differences were as follows: (1) normal faults were observed to crop out at the surface of parts of the surface trough and form fault

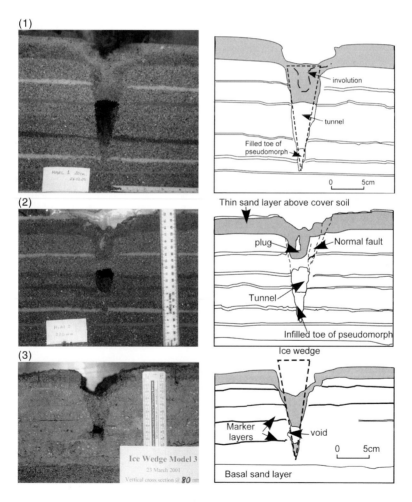

Fig. 7. Vertical sections through artificial ice-wedge pseudomorphs: (**1**) model 1, 200 mm from side of model; (**2**) model 2, 220 mm fromside of model; (**3**) model 3, 80 mm from side of model; (**4**) model 4,120 mm from side of model; (**5**) model 5, 250 mm from side of model; (**6**) model 6, 100 mm from side of model

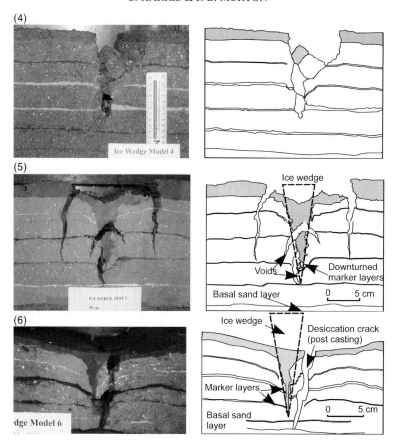

Fig. 7. Continued

scarps; (2) the involution in the centre of the upper sediment plug in some places had a flat bottom and narrow neck, and in others formed an isolated ball-and-pillow structure; (3) host sediment rather than cover soil formed the base of the plug in one vertical section (Fig. 7b), indicating that sediment adjacent to the ice wedge collapsed into a void; (4) the height of the plug was c. 23% greater (mean height = 59 ± 3 mm; 1.77 ± 0.09 m, prototype) and the height of the tunnel was c. 17% smaller (mean height = 46 ± 13 mm; 1.38 ± 0.39 m, prototype) than those in model 1. This indicates that infilling had proceeded to a greater degree in model 2.

Model 3 (Fig. 7c). The structures in model 3 broadly resembled those in models 1 and 2 (Table 2). However, several differences were noted: (1) surface cracks developed parallel, orthogonal or diagonal to the trough; (2) the sides of the trough were in places overhanging

as a result of apparent cohesion in the silt; (3) the sediment plug comprised silt derived from the cover soil, and had a distinctly looser packing than the sand plugs in models 1 and 2 and than the host silt in model 3 – the silt plug contained aggregates of typically a few millimetres in diameter, separated by voids ≤1 mm in diameter; (4) the plug generally had a wider V-shape than those in models 1 and 2 –the maximum width of the plug was on average c. 24% greater than that in models 1 and 2, and the average height of the plug was c. 38% less than that in model 1 and c. 49% less than that in model 2 (Table 2); (5) a block of cover soil tilted towards the trough centre and bounded by normal faults was observed at the top of the plug, near the trough center; (6) the maximum tunnel width was on average c. 38% less than that in model 1 and c. 31% less than that in model 2, and the tunnel height was on average c. 44% and c. 24% lower, respectively

Table 2. *Dimensions of artificial ice-wedge pseudomorphs*

	Trough width (mm)	Trough depth (mm)	Maximum plug width[a] (mm)	Plug height (mm)	Maximum tunnel width[a] (mm)	Tunnel height (mm)	Combined height of plug tunnel and toe (mm)	Maximum cast width[a] (mm)	Cast height (mm)
Model									
1	89±8 (n = 22)	35±3 (n = 22)	41±2 (n = 18)	48±3 (n = 18)	32±3 (n = 18)	63±5 (n = 18)	135±7 (n = 18)		
2	82±7 (n = 21)	35±3 (n = 21)	41±3 (n = 18)	59±3 (n = 18)	29±3 (n = 18)	46±13 (n = 18)	120±4 (n = 18)		
3	71±8 (n = 22)	42±4 (n = 22)	51±11 (n = 21)	30±5 (n = 21)	20±3 (n = 21)	35±7 (n = 21)	77±6 (n = 21)		
4	74±14 (n = 21)	68±12 (n = 21)	43±13 (n = 21)	43±27 (n = 21)	20±4 (n = 21)	38±12 (n = 21)	84±15 (n = 21)		
5	149±19 (n = 20)	30±4 (n = 20)						38±5 (n = 20)	99±12 (n = 20)
6	214±29 (n = 13)	16±1 (n = 14)						42±7 (n = 14)	84±4 (n = 14)
Equivalent prototype[b]									
1	2.67±0.24	1.05±0.09	1.23±0.06	1.44±0.09	0.96±0.09	1.89±0.15	4.05±0.21		
2	2.46±0.21	1.05±0.09	1.23±0.09	1.77±0.09	0.87±0.09	1.38±0.39	3.60±0.12		
3	2.13±0.24	1.26±0.12	1.53±0.33	0.9±0.15	0.6±0.09	1.05±0.21	2.31±0.18		
4	2.22±0.42	2.04±0.36	1.29±0.39	1.29±0.81	0.6±0.12	1.14±0.36	2.52±0.45		
5	4.47±0.57	0.90±0.12						1.14±0.15	2.97±0.36
6	6.42±0.87	0.48±0.03						1.26±0.21	2.52±0.12

Values are given to 1 standard deviation.
[a]Width of plugs, tunnels and casts were measured at right angles to the axial plane of the trough.
[b]Equivalent prototype values are ×30 model values.

(Table 2); (7) the tunnel walls were often more irregular than those in models 1 and 2, with a relief of *c.* 1–5 mm, suggestive of minor slumping; and (8) the toe was infilled with loosely packed aggregates of silt separated by voids and was often very difficult to distinguish from the host silt.

Model 4 (Fig. 7d). The trough in model 4 was almost twice as deep as that in models 1 and 2, and was shaped like a wedge with a flattened base. The sides of the trough broadly corresponded with the sides of the former ice wedge, except where a large block of host soil capped by cover soil had tilted and collapsed into the trough (Fig. 7d). The plug comprised three parts: (1) a tilted block of cover soil; (2) voids; and (3) loosely packed silty aggregates similar to those in model 3. Marker layers were generally horizontal, except beneath the large tilted block, where those beneath the side from which the block had collapsed were downturned significantly.

Model 5 (Fig. 7e). The trough in model 5 was 149 ± 19 mm wide (4.47 ± 0.57 m, prototype) and 30 ± 4 mm deep (0.90 ± 0.12 m, prototype), much wider than the troughs in models 1–4. On each side of the trough were large parallel cracks.

The ice-wedge pseudomorph in model 5 had two parts. A dominant, inner part was filled primarily with cover soil and generally well defined. A smaller, outer part comprising host soil that had subsided into the void previously occupied by the ice wedge, as indicated by downturned marker layers, was difficult to delineate precisely because the degree of downturning changed gradually with increasing distance (≤*c.* 50 mm) from the inner pseudomorph. The maximum width of the inner pseudomorph was 38 ± 5 mm (1.14 ± 0.15 m, prototype) and the height 99 ± 12 mm (2.97 ± 0.36 m, prototype). The model values represent a narrowing of 24% relative to the maximum width of the ice wedge, and a shortening of 34% relative to its height. The sides of the pseudomorph were sharp and irregular, and the toe varied from almost flat to pointed or irregular and was in places poorly defined by loose blocks of soil. The central to basal parts of the pseudomorph contained between one and several voids of irregular size and shape. The voids were often partly filled with fallen blocks of host or cover soil. Some voids extended laterally into the host soil.

Model 6 (Fig. 7f). Model 6 had a broad, saucer-like trough that was 214 ± 29 mm wide (6.42 ± 0.87 m, prototype) and 16 ± 1 mm deep (0.48 ± 0.03 m, prototype), *c.* 44% wider and *c.* 47% shallower than the trough in model 5. The trough in model 6 contained a single deep desiccation crack, rather than the two observed in model 5.

As in model 5, the ice-wedge pseudomorph in model 6 had a distinct inner part of cover soil and a less distinct outer part of host soil. The inner pseudomorph had a maximum width of 42 ± 7 mm (1.26 ± 0.21 m, prototype) and a height of 84 ± 4 mm (2.52 ± 0.12 m, prototype). The model values represent a narrowing of 17% relative to the maximum width of the ice wedge, and a shortening of 44% relative to its height. The pseudomorph had curved to irregular sides, the upper two-thirds of which were sharp and graded down into a poorly defined toe. Downturning of markers extended ≤*c.* 80 mm laterally from the inner pseudomorph and ≤*c.* 40 mm vertically. In places, the deformed markers were vertical to subvertical. A small, subvertical pipe ≤*c.* 1 mm in diameter and *c.* 1 cm long connected a rounded void within the upper part of the inner pseudomorph to an accumulation of sediment on the surface that was less stiff and lighter grey in colour than the dark grey cover soil. An involution in the form of a lobate downward protrusion of cover soil into host soil was observed several centimetres from the pseudomorph in one vertical section.

Discussion

Thaw consolidation

The amount of deformation during casting is largely determined by factors that control thaw consolidation, including the ice content of the wedge and host, the particle size of the host material, and the rate of thaw and drainage. Where ice-wedge thaw results from active-layer deepening, as in these experiments, the rate of thaw of the ice wedge tends to be slower than that of the adjacent ground. This is due to the latent heat effects of the wedge, the wedge comprising pure ice and the host comprising a mixture of soil and ice. If soil adjacent to a thawing wedge has time to consolidate, and therefore dispel much of the excess pore pressures before the adjacent wedge melts, then the soil will be much stiffer when the adjacent ice wedge melts than it would have been if the wedge melted at the same rate as the soil. Such soil is less likely to deform than soil with elev-

ated pore pressures. Sand consolidates almost instantaneously, and therefore excess pore pressures are unlikely, and the frictional soil is unlikely to deform other than by brittle failure (faulting).

As permeability decreases from sand to silt to clay, this leads to progressively slower dissipation of excess pore pressures, and slower consolidation. Potentially, therefore, the adjacent host silty clay will be softer and more deformable, especially since ice contents are usually higher. The lower coefficient of consolidation for clays means the percentage dissipation of excess pore pressures will be less in the time interval between thaw of the host soil and the adjacent ice wedge. The higher the frozen ice content of the host soil, the slower will it thaw, the greater will be the thaw consolidation, and the shorter will be the interval between soil thaw and adjacent ice-wedge thaw. Thus, pore pressures will be higher and deformation more likely in ice-rich clays, while elevated pore pressures and associated deformation become less likely as soil grain size increases and ice contents decrease.

Ice-wedge casting

Infilling of ice-wedge voids in these experiments occurred mainly by subsidence of overlying cover soil. Subsidence of cover soil in the sand and silt models (1–4) formed a sediment plug above a tunnel. Subsidence generally caused little or no disturbance of the host soil, as indicated by the absence or only limited degree of down turning of markers in models 1–4. Subsidence ceased when the plug jammed because of the downward narrowing of the wedge-shaped void. The plug was commonly supported by a natural arch along its base. Collapse of soil from the tunnel roof or walls is inferred from the loose filling of the toe and by slight irregularities on some tunnel sides. In the clayey silt (models 5 and 6), subsidence of both the cover soil and host soil produced ice-wedge pseudomorphs of composite form: an inner wedge of cover soil and an outer, poorly defined wedge of host soil that was only apparent where the marker layers were downturned.

More complete infilling of ice-wedge voids in finer grained and icier host sediments is attributed to differences in thaw consolidation and higher water contents of both host and cover soils during casting. A gradation in the extent of infilling is shown by (i) a progressive decrease in average tunnel height: 63 mm in model 1 (sand @ 15% moisture), 46 mm in model 2 (sand @ 20% moisture), 38 mm in model 4 (silt @ 20%

moisture) and 33 mm in model 3 (silt @ 40% moisture); and by (ii) the formation of an almost complete ice-wedge pseudomorph, with voids, in model 5 (clayey silt @ 30% moisture), and by a pseudomorph without voids in model 6 (clayey silt @ 60% moisture). Sand in models 1 and 2 suffered little thaw consolidation. Silt in models 3 and 4 experienced positive pore pressures during thaw consolidation, but because the rate of thaw penetration through the ice wedge was lower than through the adjacent soil, much of the soil consolidation at a given depth was complete before the adjacent ice wedge thawed. Thus when casting occurred, the host soil was sufficiently stiff to prevent excessive deformation. Clayey silt in models 5 and 6 experienced slower thaw consolidation, persistence of higher pore water pressures during thaw, and greater total thaw consolidation of the host soil. This reduced the shear strength of the thawed host soil, allowing greater deformation and infilling during casting and therefore the development of complete ice-wedge pseudomorphs.

Casting was sometimes accompanied by the formation of soft-sediment deformation and water-escape structures. Involutions in the upper, central parts of the sand plugs in models 1 and 2 took the form of lobate downward protrusions, flat bottoms and ball-and-pillow structures, consistent with an origin by load casting. Likewise, the involution observed at the base of the cover soil near the pseudomorph in model 6 is interpreted as a load cast. Involutions of similar form developed during downward thaw of ice-rich soil in previous centrifuge experiments (Harris *et al.* 2000). The pipe in model 6 is interpreted as a water-escape structure through which expelled sediment accumulated on the surface of the ice-wedge trough. The void at the base of the pipe was probably filled with a water-sediment slurry at some stage during thaw. Similar pipes formed during thaw of ice-rich clay were attributed by Harris *et al.* (2000) to water-escape events that led to rapid changes in measured porewater pressure.

Palaeoenvironmental significance

The experiments simulate, to a first approximation, ice-wedge casting associated with progressive active-layer deepening. Although natural casting is likely to be complicated by seasonal freeze–thaw processes or interrupted by renewed permafrost aggradation, some similarities between the artificial and natural ice-wedge pseudomorphs permit comment on their palaeoenvironmental significance.

Stepped normal faults in sand of model 2 (Fig. 7b) resemble graben-like faulting of sand above partially thawed ice wedges on Hooper Island, NWT, Canada (Mackay 1976), and lobate sand plugs in models 1 and 2 resemble involutions and associated downturned strata in sand and silt above partially thawed ice wedges from central Chukotka (Kotov 1986, cited in Harry & Gozdzik 1988) and western Arctic Canada (Murton & French 1993). Graben-like step faults and downturned strata in sand are also commonly reported from Pleistocene ice-wedge pseudomorphs in Europe (Dylik & Maarleveld 1967; Vandenberghe 1983), and the pseudomorphs with inner and outer parts formed in models 5 and 6 resemble the 'cone in cone' structures of some epigenetic pseudomorphs in Europe (Gozdzik 1973; Jahn 1975, pp. 69–72).

The main difference between the artificial and natural pseudomorphs concerns the plugs and tunnels formed in models 1–4. Although tunnels are commonly observed in ice wedges subject to thermal erosion, they often collapse or fill with pool ice or sediment (Mackay 1988; Murton & French 1993) and are unlikely to survive intact long after permafrost degrades. However, where voids do remain in natural pseudomorphs, they provide evidence for ice-wedge thaw (Romanovskij 1973). The models therefore suggest that casting as a result of active-layer deepening through sand and silt can be an incremental process that commences during ice-wedge thaw and finishes at some later time. The latter stages of tunnel collapse and pseudomorph formation may be unrelated to melt of ground ice. Likewise, infilling of surface troughs resulting from ice-wedge thaw, which commonly occurs in the natural environment, may post-date thaw by a considerable period of time (cf. Romanovskij 1973). Thus the final pseudomorph may often form over a variable and sometimes significant period of time.

The temperature data from our experiments indicate that thawing of the host soil occurred more rapidly than thawing of the ice wedges, which suggests that casting occurred in unfrozen soil. This disagrees with Vandenberghe's (1983) suggestion that the early stages of ice-wedge thaw induced tensile faulting in the adjacent, still-frozen ground, concurrently with sand surrounding the wedge becoming liquefied and therefore filling the space left by the melting ice. Instead, the faulting in our models and in the ice-wedge pseudomorphs described by Vandenberge more likely resulted from brittle failure in frictional, unfrozen sand in which pore water suction maintained cohesion.

A key finding from the experiments is that the degree of deformation during ice-wedge casting increased as the host sediments became finer grained and more ice-rich. In other words, the ice-wedge shape was generally best preserved in sand, and most modified in clayey silt, consistent with Dylik (1966) & Black's (1976) suggestion that ice-wedge pseudomorphs are most likely to develop and be clearly preserved in coarse-grained, ice-poor sediments. Such selective preservation has probably led to underestimation of the number of pseudomorphs in fine-grained soils that were originally ice-rich and therefore a bias towards reconstructing cold temperatures (Murton & French 1993). This is because thermal contraction cracking in sand and gravel generally requires colder ground temperatures than it does in fine-grained sediments (cf. Romanovskij 1985). Ice-wedge pseudomorphs are commonly reported from sand and gravel, and, in loessic areas in Europe, from silt. However, far fewer are reported from clays or silty clays.

Conclusions

The main conclusions from this study are as follows:

(1) Thaw consolidation played a key role in controlling sediment strength and therefore the amount of deformation and infilling that accompanied melting of ice wedges during progressive active-layer deepening.

(2) The degree of deformation during ice-wedge casting increased as the host sediments became finer grained and more ice-rich. Partially developed ice-wedge pseudomorphs formed during thaw of host sand at 15 and 20% gravimetric ice content and host silt at 20 and 40% ice content. The pseudomorphs had three parts: (a) an upper 'plug' of sediment derived largely from the cover soil; (b) a central tunnel; and (c) a basal plug of sediment that formed a 'toe' to the pseudomorph. In contrast, fully developed pseudomorphs formed during thaw of host clayey silt at 30 and 60% ice content, but thaw consolidation in the model with higher ice content led to a significant reduction in both the width and length of resulting structures. Dimensions of pseudomorphs in such host sediments may therefore not be used to infer the former size and shape of ice wedges.

(3) The experiments support the hypothesis that ice-wedge pseudomorphs tend to be better preserved in coarser-grained sediments that are not ice-rich and therefore deform little

during thaw of constituent ice wedges. This selective preservation has probably led to underestimation of the number of ice-wedge pseudomorphs in fine-grained soils that were originally ice-rich and therefore a bias towards reconstructing cold temperatures.

(4) If natural arching of coarser soils resulted during slumping from above into the ice-wedge void in a natural field setting, as observed here, it is likely that collapse of the tunnel sides would fill the void over time, widening the disturbed zone around the pseudomorph and allowing further downward collapse of soil from the surface. Thus the preservation potential of such voids is considered negligible.

This research was funded through a Royal Society of London research grant. Technical support from Mr Ian Henderson is gratefully acknowledged.

References

BLACK, R. F. 1976. Periglacial features indicative of permafrost: Ice and soil wedges. *Quaternary Research*, **6**, 3–26.

COCE, P., PANE, V., ZNIDARCIC, H., YO, H. Y, OLSEN, H. W. & SCHIFFMAN, R. L. 1985. Evaluation of consolidation theories by centrifuge modelling. *In*: CRAIG, W. H. (ed.) *Application Of Centrifuge Modelling to Geotechnical Design*, Balkema, Rotterdam, 381–401.

DYLIK, J. 1966. Problems of ice-wedge structures and frost-fissure polygons. *Biuletyn Peryglacjalny*, **15**, 241–291.

DYLIK, J. & MAARLEVELD, G. C. 1967. Frost cracks, frost fissures and related polygons. *Mededelingen van de Geologische Stichting, Nieuwe Serie*, **18**, 7–21.

FRENCH, H. M. 1996. *The Periglacial Environment*, Addison Wesley Longman, London.

GOZDZIK, J. S. 1973. Geneza I pozycja stratygraficzna struktur peryglacjalnych w srodkowej Polsce (Origin and stratigraphical position of periglacial structures in Middle Poland). *Acta Geographica Lodziensia*, **31**, 117pp.

HARRIS, C., MURTON, J. B. & DAVIES, M. C. R. 2000. Soft-sediment deformation during thawing of ice-rich frozen soils: results of scaled centrifuge modelling. *Sedimentology*, **47**, 687–700.

HARRIS, C., REA, B. & DAVIES, M. C. R. 2001. Scaled physical modelling of mass movement processes on thawing slopes. *Permafrost and Periglacial Processes*, **12**, 125–136.

HARRY, D. G. & GOZDZIK, J. S. 1988. Ice wedges: Growth, thaw transformation, and palaeoenvironmental significance. *Journal of Quaternary Science*, **3**, 39–55.

JAHN, A. 1975. *Problems of the Periglacial Zone (Zagadnienia strefy peryglacjalnef)*, Panstwowe wydawnictwo Naukowe, Warsaw.

MACKAY, J. R. 1976. Pleistocene permafrost, Hooper Island, Northwest Territories. *In*: *Current Research, Part D*, Geological Survey of Canada, Paper 76-1A, 17–18.

MACKAY, J. R. 1988. Catastrophic lake drainage, Tuktoyaktuk Peninsula area, District of Mackenzie. *In*: *Current Research, Part D*, Geological Survey of Canada, Paper 88-1D, 83–90.

MACKAY, J. R. 1990. Some observations on the growth and deformation of epigenetic, syngenetic and anti-syngenetic ice wedges. *Permafrost and Periglacial Processes*, **1**, 15–29.

MURTON, J. B. & FRENCH, H. M. 1993. Thaw modification of frost-fissure wedges, Richards Island, Pleistocene Mackenzie Delta, western Canadian Arctic. *Journal of Quaternary Science*, **8**, 185–196.

MURTON, J. B. & HARRIS, C. 2003. The experimental simulation of ice-wedge casting. *In*: PHILLIPS, M., SPRINGMAN, S. M. & ARENSON, L. U. (eds) *Proc. 8th International Conference on Permafrost*, 21–25 July 2003, Zurich. Swets and Zeitlinger, Lisse, 807–810.

MURTON, J. B. & KOLSTRUP, E. 2003. Ice-wedge casts as indicators of palaeotemperature: Precise proxy or wishful thinking? *Progress in Physical Geography*, **27**, 155–170.

ROMANOVSKIJ, N. N. 1973. Regularities in formation of frost-fissures and development of frost-fissure polygons. *Biuletyn Peryglacjalny*, **23**, 237–277.

ROMANOVSKIJ, N. N. 1985. Distribution of recently active ice and soil wedges in the U.S.S.R. *In*: CHURCH, M. & SLAYMAKER, O. (eds) *Field and Theory: Lectures in Geocryology*, University of British Columbia Press, Vancouver, 154–165.

SAVIDOU, C. 1988. Centrifuge modelling of heat transfer in soil. *Centrifuge*, **88**, 583–591.

VANDENBERGHE, J. 1983. Ice-wedge casts and involutions as permafrost indicators and their stratigraphic position in the Weichselian. *In*: *Permafrost: Fourth International Conference, Proc.*, Fairbanks, AK, National Academy Press, Washington, DC, 1298–1302.

Late Holocene solifluction history reconstructed using tephrochronology

MARTIN P. KIRKBRIDE[1] & ANDREW J. DUGMORE[2]

[1]*Department of Geography, University of Dundee, Dundee DD1 4HN, Scotland, UK*
(e-mail: m.p.kirkbride@dundee.ac.uk)
[2]*School of GeoSciences, University of Edinburgh, Drummond Street,*
Edinburgh EH8 9XP, Scotland, UK
(e-mail: ajd@geo.ed.ac.uk)

Abstract: Phases of activity of four solifluction lobes at an altitude of 750–800 m are dated by tephrochronology at Snaefell, central eastern Iceland (64°48′ N 15°33′ E). The sample includes sorted lobes with tread gradients of 3–11° and unsorted (turf-banked terraces) in slope-foot locations. Trenches through lobe fronts reveal detailed internal structures picked out by multiple tephra layers. The tephras V1717, V1477, Ö1362, V870, Hekla-3 (2900 years BP) and Hekla-4 (3800 years BP) provide isochronous surfaces of known age whose deformation and disturbance indicate mass movement and/or cryoturbation of the soil cover. Undisturbed soil including the Hekla-3 tephra indicates an absence of solifluction prior to 2900 years BP. Several centuries after Hekla-3, gravel-rich horizons mark widespread frost heave and solifluction of hillslopes. Later stabilization of these lobes allowed the accumulation of aprons of aeolian sediment below lee-side risers. These aprons contain *in situ* mediaeval tephras, dating the inception of solifluction to a considerable time prior to Norse settlement. The likely period of this first phase of solifluction is the Later Bog Period of the Subatlantic, *c.* 2500–1000 years BP. The aprons are currently being overridden and deformed by solifluction lobes reactivated in the Little Ice Age.

This paper presents results of a study of solifluction at 750–850 m altitude on the northeastern slopes of Snaefell, in east-central Iceland (Fig. 1). The aim is to reconstruct the timing of periods of active solifluction and to interpret active phases in relation to known climate changes and anthropogenic impacts. Dating control is provided by six tephra layers of known provenance and age. Elsewhere, ^{14}C dating of organic material beneath solifluction lobes has provided chronologies of activity (Benedict 1970; Worsley & Harris 1974; Matthews *et al.* 1986), although such dating is fraught with methodological problems (Matthews *et al.* 1986). Hirakawa (1989) pioneered the dating of solifluction in Iceland using *in situ* tephra layers. From detailed study of two sites, he concluded that individual lobes contain an archive of motion covering at least the last 7000 years. Reconstructed long-term rates of activity were greatest between the Hekla-4 and Landnam tephras (*c.* 3800–1100 years BP), when rates were at least double those of the last 900 years. The potential for further developing a precise tephra-based chronology has not subsequently been realized.

The climatic history of Iceland in the Holocene has been reconstructed primarily from stratigraphic and palynological analyses of peats (Einarsson 1963; Vasari 1972; Hallsdóttir 1995; Wastl *et al.* 2001) and dating of terrestrial glacial deposits (Guðmundsson 1997; Kirkbride & Dugmore 2001a). Major ecological change occurred in the Subatlantic period *c.* 2000–2500 years BP. Loss of birch forest was accompanied by the spread of heath and bog vegetation, heralding the so-called Late Bog Period from 2500 to 1000 years BP (Einarsson 1961). Glaciers advanced on several occasions since *c.* 5000 years BP, during the Neoglacial period, most recently in the Little Ice Age (LIA) of the last *c.* 800 years (Grove 1988; Dugmore 1989; Evans *et al.* 1999; Kirkbride & Dugmore 2001b). Norse settlement at *c.* AD 870–920 heralded a period of widespread anthropogenic destabilization of the landscape, manifested as aeolian and waterborne soil erosion and exposure of rock pavements in denuded areas (Buckland *et al.* 1991; Dugmore & Erskine 1994; Arnalds 1997; Dugmore *et al.* 2000; Olafsdóttir & Guðmundsson 2002) and the formation of frost

From: HARRIS, C. & MURTON, J. B. (eds) 2005. *Cryospheric Systems: Glaciers and Permafrost.*
Geological Society, London, Special Publications, **242**, 145–155.
0305-8719/05/$15.00 © The Geological Society of London 2005.

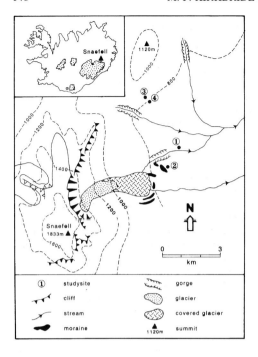

Fig. 1. Location map of the study site at Snaefell, showing the cirque glacier Hálsajökull and the locations of excavated profiles in four solifluction lobes

(4) the cooler centuries of the LIA commencing in the fourteenth century AD and following the Mediaeval Warm Period (MWP) of the tenth to thirteenth centuries AD (Grove 1988).

The timing of phases of activation of solifluction is significant, because they potentially show whether glacial and periglacial systems are in phase, both driven by temperature change; whether ecological and periglacial changes are coevally associated and driven by precipitation (acting through greater soil wetness and seasonal ground ice segregation); or whether later anthropogenic changes to vegetation cover and deteriorating climate combined to trigger solifluction for similar reasons.

Tephra layers are employed in three ways in this study. First, those tephras whose age is known are used as isochrones to date the contemporary soil surface. Second, tephra layers define the sedimentary bedding of their host sediment, aiding facies analysis. Third, distinctive patterns of post-depositional disturbance of tephra layers allow interpretations to be made of the processes acting on the soil, notably folding and faulting associated with the motion of solifluction sheets, and partial disruption of the layers by frost heave. Thus, considerably more morphostratigraphic information is available to the researcher in this environment than would be the case in other circumstances, where ^{14}C dating has been the main chronological tool employed. This lacks the resolution to distinguish between late mediaeval and LIA events, and relationships between the dates obtained and geomorphic events are not always clear. The visual power of tephrostratigraphy is of great benefit for the field-testing of working hypotheses.

Solifluction forms in the study area

The lower eastern slopes of Snaefell (Fig. 1) are mantled by solifluction sheets whose crenulate risers often overlap to form individual lobes. Some parts of the slope contain wholly vegetated features at present (non-sorted sheets and lobes, *sensu* Ballantyne & Harris 1994), but elsewhere the treads form pavements of sparsely vegetated frost-sorted soil between well-vegetated risers (sorted sheets and lobes or turf-banked terraces).

The risers are heavily vegetated by at least 20 plant species, including mosses, sedges, heath plants such as *Empetrum nigrum*, *Alchemilla alpina*, *Vaccinium uliginosum*, and shrubs including *Betula nana*, *Juniperus communis*, *Salix herbacea* and *Salix lanata*. The characteristic

hummocks (thufur) on grazed and cultivated land (Webb 1972). Anthropogenic degradation was greatly exacerbated by the climatic severity of the middle and later LIA, particularly after the sixteenth century AD, when reduced growing seasons and fodder shortages led to increased grazing pressure on rangelands.

It is within this environmental history that the timing of active solifluction will be investigated. In particular, evidence is sought for the activation of solifluction, specifically with reference to four significant climate/environmental changes:

(1) the earliest mid-Holocene (Neoglacial) glacier expansion, culminating between 5000 BP and 4000 years BP, driven primarily by decreasing temperature;

(2) ecological changes after *c.* 2500 years BP, corresponding to Einarsson's (1961) Late Bog Period (LBP), driven primarily by increasing precipitation;

(3) land cover change following rapidly from human settlement around AD 870, heralding the clearing of scrub and woodland and the introduction of grazing animals;

riser/tread morphometry of sorted forms is summarized by correlations between overall slope angle with riser heights and tread lengths respectively, for 19 individual lobes (Fig. 2). Maximum riser angles exceed 50° except where risers have been degraded by localized slumping, and riser heights are mostly within the range 0.2–1.5 m. Tread lengths are typically in the range 10–30 m. A subordinate scale of feature is often superimposed on the main solifluction sheets. These smaller features have treads and risers of c. 1 and c. 0.1 m respectively, and are common on larger treads which steepen towards the main riser. Vegetation cover extends downslope from most risers onto the subjacent tread, usually forming an apron of slightly steeper angle extending from 1 to c. 5 m along the upslope fringe of the lower tread. The downslope margin of the vegetated apron is often marked by a zone of incomplete vegetation, forming short oblique stripes interspersed with bare ground. In places, metre-scale lobes of clasts spill over the vegetated risers (see below).

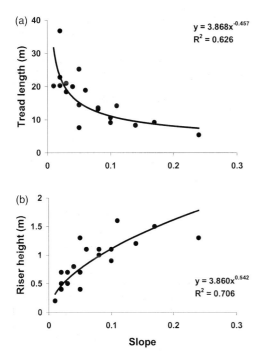

Fig. 2. Morphometric relationships between (**a**) tread length and slope angle, and (**b**) riser height and slope angle measured over the length of the solifluction lobe. 19 individual riser-tread pairs were measured up- and down-slope from sites 3 and 4 in Fig. 1

Research methods

Trenches were excavated through four risers to examine the internal stratigraphy of the solifluction sheets. Site 1 was a fully vegetated (non-sorted) solifluction lobe, cut through by the northern outwash stream from the nearby cirque glacier Hálsajökull (Fig. 3a). Site 2 was a non-sorted lobe excavated on the flank of a vegetated moraine ridge close to the terminus of Hálsajökull (Fig. 3b). Sites 3 and 4 were the risers of adjacent sorted solifluction sheets several hundred metres upslope to the north (Fig. 3c). In all cases, trenching exposed a cross section of the active front and the subjacent sediment. At the sorted lobe fronts, vegetated aprons below the risers were excavated to their downslope fringe. Sections were photographed, sketched and logged, and detailed stratigraphic profiles recorded.

Tephrostratigraphy is based on the field identification of the tephra layers in soil profiles, using visible macroscopic features of tephras (Table 1). Samples were collected of key marker layers V870 and Hekla-3, and analysed by electron microprobe to confirm their field identification, using the protocols outlined by Larsen *et al.* (1999, 2001).

Results

Tephrostratigraphy

Frequent volcanic activity has deposited many airfall tephras, forming visible markers within aeolian soils mantling hillslopes and peat soils on valley floors. Tephra layers form regional isochrones which conveniently subdivide the second half of the Holocene (Table 1). The V1717 (Veidivotn AD 1717) tephra is found only where vegetation cover has been present since the eruption date. Although thin and sometimes patchy in distribution, the tephra is nevertheless a useful marker for the mid-LIA. The Ö1362 (Öraefajökull AD 1362) and Layer 'a'/V1477 (Veidivotn AD 1477) are particularly distinct, and provide marker isochrones closely predating the colder centuries of the LIA (Grove 1988; Ogilvie 1992). The basic (Veidivotn) component of the Landnam tephra V870 (c. AD 870), marks the land surface close to the time of human settlement (Dugmore *et al.* 2000). Together, V870 and Ö1362 conveniently bracket the MWP. Hekla-3 (c. 2900 years BP) was not identified in the field but its presence in a monolith M3 recovered from site 1 was later confirmed by geochemical analysis (Fig. 3a). Hekla-4 (c. 3800 years BP)

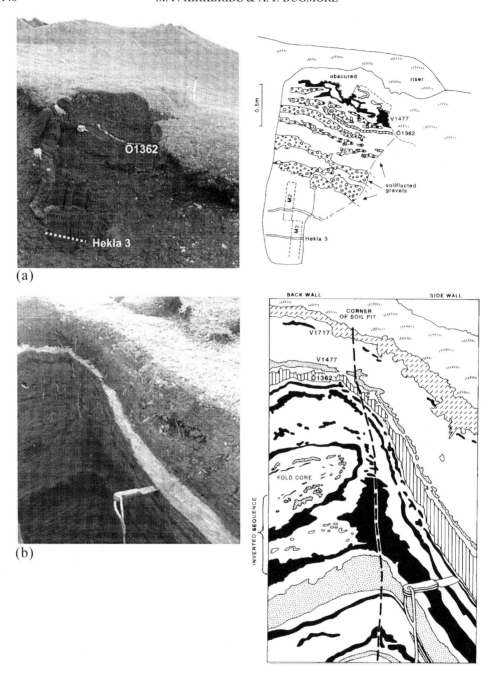

Fig. 3. Photographs and field sketches of three of the trenched features. (**a**) The stream-cut section through the riser of a currently non-sorted solifluction sheet, site 1. The prominent white tephra layer is Öraefajökull AD 1362 (Ö1362). Note the gravel layers above the Hekla-3 tephra. M2 and M3 refer to monoliths extracted from the section for detailed analysis. (**b**) Soil profile at site 2 on an old terminal moraine of Hálsajökull, showing the buried riser of a non-sorted lobe picked out by rotation, inversion, and hinge-zone thickening of tephra layers. Note how the Ö1362 tephra and the superjacent layers drape passively over the inactive solifluction lobe. (**c**) Trench through a vegetated apron forming the upper part of a tread at site 3. The riser above has been trenched and back-filled. Note the tapering soil thickness in the trench wall: see Fig. 6d for details. Active frost-heaved clast pavements are visible below the apron and above the riser

(c)

Fig. 3. Continued

is useful for separating the earliest Neoglacial period from later periods of time, and Hekla-3 for subdividing the Late Bog Period from earlier neoglacial events. Numerous unidentified thin basaltic tephras occur between these known layers. They provide relative age control and reveal bedding and deformational structures, but do not provide absolute ages.

Sediment facies within soliflucted soil: sorted forms

Five sediment facies were identified within the soliflucted soil, distinguished by their lithology and/or internal structure revealed by the enclosed tephra layers. Spatial relations between the facies are summarized in Fig. 4.

Facies 1: undisturbed interlayered soil and tephra. Facies 1 results from background aeolian deposition, which is the primary source of soil parent material in inland Iceland over the Holocene. Texturally, aeolian soils are predominantly well sorted orange-brown to yellow-brown silt and sandy silt, often with variable admixtures of coarser tephra grains, especially above major tephra layers. Long-term, uniform soil accumulation depends on the presence of a dense vegetation cover to trap aeolian particles. Ongoing accumulation, compaction and decay of the organic component allow preservation of fine layers within the aggrading sequence. Typical prehistoric accumulation rates lie within the range 0.1–0.3 mm/a (Dugmore & Erskine 1994; Olafsdóttir & Guðmundsson 2002). In historic time, widespread destabilization of aeolian soils in Iceland has given rise to accumulation rates in receiving areas of ≤1–2 mm/year,

Table 1. *Tephra layers used as stratigraphic markers on Snaefell hillslopes*

Layer	Age	Texture	Colour
Veidivotn V1717	AD 1717	Fine	Black
Veidivotn V1477	AD 1477	Fine	Olive-green to grey
Öraefajökull Ö1362	AD 1362	Fine	White
Landnam V870	c. AD 870	Fine	Olive-green with light crystals
Hekla 3	2900 years BP	Fine	White
Hekla 4	3800 years BP	Couplet comprising:	
		(upper) fine	Black
		(lower) fine	White
Hekla 5	6000 years BP	Fine	White

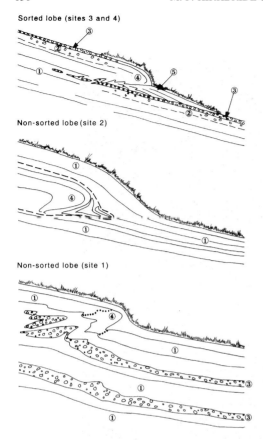

Fig. 4. Diagram showing the essential spatial relations between facies for sorted and non-sorted lobes. The numbers refer to the facies described in the text. Facies 1: undisturbed interlayered soil and tephra. Facies 2: massive or poorly layered soil/tephra. Facies 3: soliflucted clast-rich soil. Facies 4: layered soil and tephra bound by turf mat on riser slopes. Facies 5: layered soil and tephra forming aprons below risers

particularly during the severe climates since the early eighteenth century AD (Dugmore *et al.* 2000). Generally, bedding in facies 1 lies parallel to underlying surfaces where these are visible (Fig. 5).

Facies 2: massive or poorly layered soil/ tephra. Facies 2 is derived from facies 1 by cryoturbation, in which most or all of the tephra stratigraphy has been destroyed by heave. Tephra layers are reduced to dispersed grains, but occasionally lenses and 'smudges' are preserved with sufficient coherence to allow identification of the tephra and its approximate location within a profile. White tephras such as Ö1362 remain more visible than dark tephras such as V1477. Facies 2 forms to a depth of 25–40 cm under presently devegetated surfaces subject to seasonal frost action. Facies 2 has not been observed beneath well-vegetated surfaces, indicating that the anchoring and insulation provided by the turf mat suppress frost heave in the soil beneath.

Facies 3: soliflucted clast-rich soil. Soliflucted material comprises massive admixtures of aeolian silt and silty sand, clasts and dispersed tephra. The layer may be unsorted, or may include a clast pavement or clast-rich lenses. This facies occurs both as the surface layer at active sites and as buried layers or lenses (Fig. 5). Several downslope trends are evident within individual tread surfaces (Fig. 6). These include increasing concentration of clast cover, increasing stability of surface clasts shown by weathering and lichen cover, and occasional areas of decimetre-scale sorted stripes or nets merging downslope into areas of greater clast concentration. The much higher downslope mobility of the frost-heaved coarse fraction compared with the fine matrix is shown by clast lobes spilling down risers (Fig. 6c). Rapid downslope motion of a shallow surface layer is also indicated by the alignment of the tap-roots of moss campion (*Silene acaulis*), exposed by heave and transport of the plant cushion away from the deeply anchored root. Excavations show buried clast-rich layers of facies 3 beneath facies 1 and 4 soils, indicating their potential for burial and preservation within aggrading soils (see Figs 3 and 5).

Understanding the origin of the clasts in facies 3 is important for interpreting the dynamics of the slope system as a whole. The primary aeolian soil is clast-free, so the presence of clasts is evidence of transport of a mobile carpet of frost-heaved surface clasts over considerable distances, and probably at considerable (but unknown) rates. Evidence suggests that clasts are derived from bedrock outcrops several hundred metres upslope and have been transported in a rather thin near-surface layer of very active frost heave, suggesting needle ice as the mechanism (cf. Higashi & Corte 1971). It is possible that most of the mass transport on the slope is of this type, as shown by the high mobility of clast lobes overriding riser slopes (Fig. 6c). Downslope clast velocities cannot be accurately reconstructed, but must be at least an order of magnitude greater than the advance of the solifluction lobes themselves based on the distance from upslope rock outcrops and time constraints. The alternative hypothesis, that clasts have been up-heaved through the whole profile from the local rockhead, is

Fig. 5. Soil profile at site 1, a few metres downstream from the main section shown in Fig. 3a. Undisturbed soil and tephra (facies 1) lies above and below a clast-rich soliflucted layer of facies 3

discounted because of the thickness of the soil profiles, the preservation of stratigraphy at depth (Figs 3a and 5), and the absence of clast-rich layers in nearby soil profiles where solifluction has not occurred.

Facies 4: layered soil and tephra bound by turf mat on riser slopes. Terrace and lobe risers are invariably vegetated, and therefore trap aeolian silt even when treads up- and downslope are devegetated and subject to frost heave. Tephra layers (notably V1717) are visible within actively deforming vegetated risers, which are therefore not subject to disruption of the stratigraphy. Even where soil is overturned as the advancing solifluction lobe 'rolls over', the stratigraphy can be preserved as small-scale inversions. The presence of deformed bedding distinguishes this facies from facies 1. In both the trenched sorted lobes, tephra layers pick

out recumbent folds with axial planes dipping upslope and hinges comprising undisturbed tephra in the apron (facies 5) contiguous with over-folded tephra in the basal part of the overriding lobe. These structures contrast with the parallel bedding of facies 1 and the onlapping bedding of facies 5.

Facies 5: layered soil and tephra forming aprons below risers. The re-entrant between solifluction risers and treads forms a sheltered location favouring vegetation growth, and therefore a lee-side trap for aeolian soil. Facies 5 represents accumulation of soil and tephra in such locations, forming an apron or wedge tapering downslope (Fig. 6d). Bedding in the apron dips gently downslope, parallel to the present ground surface but onlapping over the adjacent tread. The facies is significant in that: (1) it post-dates formation of the solifluction riser—the lowest onlapping

Fig. 6. Characteristics of soliflucted clast-rich layers. (**a**) Incipient stone stripes on upslope part of a tread surface. (**b**) Relatively stable clast pavement with larger lichen-bearing clasts, sporadic vegetation, and weakly developed sorted nets with fine-grained centres (under tape measure). Tape extends 30 cm. (**c**) Mobile clast pavement spilling down the riser of a solifluction sheet near sites 3 and 4. Spade is positioned towards the top of the apron of facies 5. (**d**) Detail of facies 5 overlying clast-rich facies 3. Tephra layers 'a'/V1477 and Ö1362 are marked (see knife for scale)

tephra lies closest to the re-entrant, and provides the minimum age of the inception of the landform; (2) it indicates a slow, but not necessarily continuous, downslope colonisation by vegetation and soil accumulation on tread surfaces since the MWP.

Facies relationships in space and time: sorted forms

Fieldwork has provided a snapshot of the present spatial disposition of facies. It is recognized that the observed facies distribution is almost certainly transient, and results from the spatial and temporal variability of several concurrent processes. Interpretation of facies relationships (Fig. 4) requires a dynamic interpretation involving vertical accretion of soil on vegetated surfaces (which creates the stratigraphy), perturbation of soil by frost action on unvegetated surfaces during cooler climates and downslope

motion of the soliflucting soil, which will attenuate the profile in upslope locations, thicken it downslope and cause local folding and overturning at active risers.

The presence and role of vegetation is crucial, because it both inhibits cryoturbation and allows ongoing soil and tephra accumulation. Heathland vegetation mats ubiquitously preserve fine tepha layers unless airfall occurs onto a snowpack. Wherever a tephrostratigraphy is preserved, it follows that the vegetation cover has been present for at least as long as it takes to aggrade a depth of soil equal to the depth of seasonal frost penetration ($c. \geq 0.25$ m, representing a few centuries in post-settlement time).

Sediment facies within soliflucted soil: non-sorted forms

Presently non-sorted solifluction terraces mostly occupy slope foot locations and currently sustain a complete and lush vegetation cover over peaty soils, in contrast to the bare tread surfaces over aeolian soils on the sorted forms. At site 1 (Fig. 3a), a complete mid-Holocene sequence of undisturbed soil (facies 1) is overlain by disturbed soils and clast-rich layers of facies 1, 3 and 4. Facies 2 is suppressed and facies 5 not observed. The presence of three massive clast-rich layers above the Hekla-3 tephra (Fig. 3a) suggests several phases of more active frost heave and solifluction with reduced vegetation cover after 2900 years BP, and the formation of sorted forms unlike the present vegetated forms. The section through the present riser demonstrates that solifluction during the LIA overrides the Ö1362 and V1477 tephras, which pick out drag folds and dislocations beneath the active lobe. The thicker peat profile in this low-gradient slope-foot location means that soil accumulation has been more significant than downslope displacement of the soil mass. Located at the break of slope, this site appears to record stacking of gravel-rich soliflucted layers in the 'run-out zone' of mass transport from upslope, but only after deposition of the Hekla-3 tephra. Significantly, the history of activity revealed at this site concords with the sorted lobes in thinner aeolian soils in the upslope location.

At site 2, a buried, inactive lobe front has been draped by tephras including Ö1362 which show no subsequent deformation due to a very local soil reservoir (the moraine crest), providing only a single phase of activity (Fig. 3b). The V870 tephra appears to have been deformed by solifluction, including thickening in the hinge of the overfold. Undisturbed tephras below V870 indicate a later inception of solifluction at this site, probably around the seventh to ninth centuries AD, on a moraine deposited in the Sub-atlantic period. No LIA activity is recorded at this site.

The history of slope destabilisation

A reconstruction of the development of the sorted solifluction lobes is made in Fig. 7. There is no evidence for destabilization of slopes by solifluction prior to deposition of the Hekla-3 tephra at c. 2900 years BP, either at the trenched lobe sites or at other soil profiles elsewhere in the area. We note that evidence for solifluction as early as c. 7000 years BP in western Iceland (Hirakawa 1989) was only revealed by trenching up to 12 m upslope from lobe risers. However, it is possible that older tephras at these sites have been deformed by drag-folding driven by solifluction in superjacent soil, and

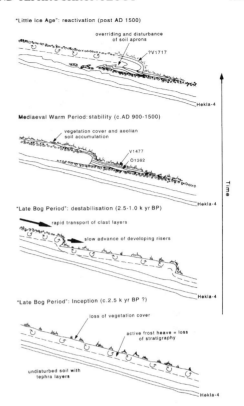

Fig. 7. Reconstruction of the history of solifluction at Snaefell since the Late Bog Period (bottom picture) to the LIA (top picture). Clasts transported by frost heave from upslope rock outcrops progressively mantle the slope as vegetation cover becomes sparse and solifluction gives rise to tread-and-riser topography. Stabilization in the Mediaeval Warm Period associated with recovery of vegetation allows aprons of soil and tephra to accumulate in the downslope lee of risers, and tephra layers to be trapped in thin undisturbed soil on tread surfaces. Finally, reactivation in the LIA is associated with devegetation of tread surfaces and disturbance of their mediaeval soils, and causes risers to override their downslope aprons

therefore may overestimate the timing of solifluction.

At Snaefell, it is not possible to date precisely the inception of solifluction because cryoturbation has destroyed the soil stratigraphy in facies 2 immediately below the present tread surface. It is possible to conclude that the earliest frost-heaved clast-rich layers and soliflucted soil exposed in trenches (facies 3) formed after c. 2900 years BP and some considerable time prior to deposition of the Ö1362 tephra within facies 5 aprons. Given that the MWP extended from the tenth to the thirteenth centuries, it is likely that the inception of cryoturbation also predates mediaeval times. Crucial to this interpretation is the presence of mediaeval tephra layers within soils of facies 5 which *overlie* a clast-rich layer, demonstrating that frost heave predated the Ö1362 tephra. The Ö1362 tephra also occurs as patches in facies 2, indicative of contemporary vegetation cover and soil accumulation on what is now a bare frost-heaved surface. This shows that, during the MWP, vegetation colonized a previously cryoturbated surface, which subsequently lost its vegetation cover, allowing cryoturbation to resume, to mix the Ö1362 tephra within the soil beneath the clast pavement.

This reasoning implies that solifluction probably commenced during what Einarsson (1961) termed the Late Bog Period, a deterioration of climate across Iceland and northern Europe at around 2500 years BP which occurred within the Subatlantic period of the palynological record. Cooler and wetter climates were associated at the study site with wholesale loss of vegetation cover, to expose the soil mantle to severe seasonal freezing and thawing. Frost heave and sorting concentrated clasts at the surface and destroyed layering in the upper *c.* 0.25 m of soil, possibly causing the 'disappearance' of the Hekla-3 tephra, which would have occurred as a thin (<10 mm) layer within a few centimetres of the contemporary soil surface. Solifluction of this disturbed layer is interpreted to have commenced at this time.

During the MWP, the slope was stabilized by vegetation, retarding or possibly preventing solifluction and allowing aeolian soil to accumulate across the whole slope, as shown by the presence of traces of the Ö1362 tephra within facies 2 at sites 3 and 4. Only in the lee-side aprons of facies 5 has bedding been preserved. As the climate became more severe in the LIA, solifluction lobes which had formed before the MWP were reactivated, causing overriding of the upslope parts of the facies 5 aprons and overfolding of the tephra-bearing soil which had accumulated in the vegetated risers during mediaeval times. Cryoturbation in exposed, devegetated treads reduced the Ö1362 tephra to a few patchy traces beneath the surface pavement. In contrast, the continued survival of vegetation on risers and aprons is shown by the preservation of the LIA tephras V1477 and V1717.

In summary, periods of stability were characterized by increased vegetation cover, trapping of aeolian silt and tephra, suppressed frost-action and subduing of the slope microrelief. Periods of activity were characterized by reduced vegetation cover, increased frost penetration, and very rapid

downslope heave of surface clasts across the actively soliflucting soil mass.

Conclusions

The study allows the following specific conclusions to be drawn:

(1) If the Snaefell slopes are representative of inland eastern Iceland, the inception of solifluction in the region probably occurred in the Late Bog Period (the Subatlantic) in the third millennium BP.

(2) The lack of contemporary disturbance to mediaeval-age tephras within aeolian soil indicates that the milder climate of the MWP was associated with a period of stabilization of soliflucted slopes.

(3) Re-activation during the subsequent LIA led to overriding and deformation of mediaeval-age sediments. In the absence of tephrochronological dating, it would have been easy to misinterpret *all* the solifluction activity in the area as a consequence of the LIA.

(4) Both sorted forms on exposed hillslopes and unsorted forms in vegetated slope-foot locations reveal similar histories of activity. However, clast-rich horizons within presently unsorted solifluction sheets suggest that reduced vegetation cover and frost-sorting have affected even the lower slopes in the past.

More generally, these findings lead to the conclusion that the periglacial slope system responded to climate deterioration some time after the initial 'neoglacial' advance of glaciers (cf. Guðmundsson 1997; Kirkbride & Dugmore 2001a). Renewed solifluction during the LIA does broadly correlate with glacial advances, but only at a coarse timescale. Solifluction appears to be more closely associated, probably causatively, with climatically driven changes in vegetation cover. There is apparently no geomorphological signal of anthropogenic land cover change due to the introduction of rangeland grazing, although such land cover changes occurred in the tenth century during a period of relatively equable climate.

The authors acknowledge funding from the Carnegie Trust for the Universities of Scotland, and the Leverhulme Trust. The enthusiastic field assistance of Matthew Bull and Andrew Caseley in 2002 is appreciated. Two anonymous referees, whose perceptive comments improved the clarity of the manuscript, are gratefully acknowledged.

References

ARNALDS, A. 1987. Ecosystem disturbance and recovery in Iceland. *Arctic and Alpine Research*, **19**, 508–513.

BALLANTYNE, C. K. & HARRIS, C. 1994. *The Periglaciation of Great Britain*, Cambridge University Press, Cambridge.

BENEDICT, J. B. 1970. Downslope movement in a Colorado alpine region: rates, processes and climatic significance. *Arctic and Alpine Research*, **2**, 165–226.

BUCKLAND, P. C., DUGMORE, A. J., PERRY, D. W., SAVOURY, D. & SVEINBJARNARDOTTIR, G. 1991. Holt in Eyjafjallasveit, Iceland. A palaeoecological study of the impact of landnam. *Acta Archaeologica*, **61**, 252–271.

DUGMORE, A. J. 1989. Tephrochronological studies of Holocene glacier fluctuations in south Iceland. *In*: OERLEMANS, J. (ed.) *Glacier Fluctuations and Climatic Change*, Kluwer Academic, Dordrecht, 37–55.

DUGMORE, A. J. & BUCKLAND, P. C. 1991. Tephrochronology and late Holocene soil erosion in Iceland. *In*: MAIZELS, J. & CASELDINE, C. (eds) *Environmental Change in Iceland*, Kluwer Academic, Dordrecht, 147–159.

DUGMORE, A. J. & ERSKINE, C. C. 1994. Local and regional patterns of soil erosion in southern Iceland. *Münchener Geographische Abhandlungen*, **12**, 63–79.

DUGMORE, A. J., NEWTON, A. J., LARSEN, G. & COOK, G. T. 2000. Tephrochronology, environmental change, and the Norse settlement of Iceland. *Environmental Archaeology*, **5**, 21–34.

EINARSSON, P. 1961. Pollenanalyische untersuchung zur spät- und postglazialen klimageschichte Islands. *Sönderveröffentischungen des Geologischen Institutes der Universitat Koln*, **6**, 1–52.

EINARSSON, P. 1963. Pollen analytical studies on the vegetation and climate history of Iceland in Late and Postglacial times. *In*: LÖVE, A. & LÖVE, D. (eds) *North Atlantic Biota and their History*, Pergamon Press, Oxford, 355–365.

EVANS, D. J. A., ARCHER, S. & WILSON, D. J. H. 1999. A comparison of the lichenometric and Schmidt hammer dating techniques based on data from the proglacial areas of some Icelandic glaciers. *Quaternary Science Reviews*, **18**, 13–41.

GROVE, J. M. 1988. *The Little Ice Age*, Methuen, London.

GUÐMUNDSSON, H. J. 1997. A review of the Holocene environmental history of Iceland. *Quaternary Science Reviews*, **16**, 81–92.

HALLSDÓTTIR, M. 1995. On the pre-settlement history of Icelandic vegetation. *Icelandic Agricultural Sciences*, **9**, 17–29.

HIGASHI, A. & CORTE, A. E. 1971. Solifluction: A model experiment. *Science*, **171**, 480–482.

HIRAKAWA, K. 1989. Downslope movement of solifluction lobes in Iceland: A tephrostratigraphic approach. *Tokyo Metropolitan University Geographical Reports*, **24**, 15–30.

KIRKBRIDE, M. P. & DUGMORE, A. J. 2001a. Timing and significance of mid-Holocene glacier advances in northern and central Iceland. *Journal of Quaternary Science*, **16**, 145–153.

KIRKBRIDE, M. P. & DUGMORE, A. J. 2001b. Can lichenometry be used to date the 'Little Ice Age' glacial maximum in Iceland? *Climatic Change*, **48**, 151–167.

LARSEN, G., DUGMORE, A. J. & NEWTON, A. J. 1999. Geochemistry of historical age silicic tephras in Iceland. *The Holocene*, **9**, 463–471.

LARSEN, G., NEWTON, A. J., DUGMORE, A. J. & VILMUNDARDÓTTIR, E. 2001. Geochemistry, dispersal, volumes and chronology of Holocene silicic tephra layers from the Katla volcanic system. *Journal of Quaternary Science*, **16**, 119–132.

MATTHEWS, J. A., HARRIS, C. & BALLANTYNE, C. K. 1986. Studies on a gelifluction lobe, Jotunheimen, Norway: ^{14}C chronology, stratigraphy, sedimentology and palaeoenvironment. *Geografiska Annaler*, **66A**, 345–360.

OGILVIE, A. E. J. 1992. Documentary evidence for changes in the climate of Iceland, A.D. 1500 to 1800. *In*: BRADLEY, R. S. & JONES, P. D. (eds) *Climate Since A.D. 1500*, Routledge, London, 92–117.

OLAFSDÓTTIR, R. & GUÐMUNDSSON, H. J. 2002. Holocene land degradation and climatic change in northeastern Iceland. *The Holocene*, **12**, 159–167.

VASARI, Y. 1972. The history of the vegetation of Iceland during the Holocene. *In*: HYVARINEN, H. & HICKS, S. (eds) *Climatic Changes in the Arctic during the last Ten Thousand years*. Acta Universitatis Ouluensis, 239–251.

WASTL, M., STÖTTER, J. & CASELDINE, C. J. 2001. Reconstruction of the Holocene variations of the upper limit of tree or shrub birch growth in northern Iceland based on evidence from Vesturárdalur-Skíðadalur, Trollaskagi. *Arctic, Antarctic and Alpine Research*, **33**, 191–203.

WEBB, R. 1972. Vegetation cover on Icelandic thufur. *Acta Botanica Islandica*, **1**, 51–60.

WORSLEY, P. & HARRIS, C. 1974. Evidence for Neoglacial solifluction at Okstindan, north Norway. *Arctic*, **27**, 128–144.

Index

Note: Page numbers *in italics* refer to figures; those **in bold** refer to tables.